An Engineering Foundation Conference

Foundations for
DAMS

Asilomar Conference Grounds
Pacific Grove, California
March 17-21, 1974

Co-Sponsored by
Geotechnical Engineering Division, ASCE
United States Committee on Large Dams

AMERICAN
SOCIETY OF
CIVIL
ENGINEERS
FOUNDED
1852

Published by
American Society of Civil Engineers
345 East 47th Street
New York, N.Y. 10017

Price $12.00

CONTENTS

iii

FOREIGN EXPERIENCE: DAMS ON DIFFICULT SITES

FOUNDATION MAINTENANCE: REPAIRS AND SPECIAL PROBLEMS

ENGINEERS AND GEOLOGISTS

*Manuscript not available

CLOSING SESSION

INDEXES

FOREWORD

The Engineering Foundation's Conference on "Foundations for Dams" held in March 1974 at Asilomar, Pacific Grove, California, attracted over 80 engineers and geologists. Five foreign countries were represented among the speakers. The talks and discussions centered around difficult foundation problems and foundation grouting. Both subjects are of growing interest as the so-called "good sites" are becoming more scarce and the needs for larger dams are increasing.

Construction of concrete structures (spillways, powerhouses, or navigation locks) on soft foundations ranged from floating them directly on river alluvium or sand deposits with partial sheet pile cut-offs to supporting them on steel piles. The examples of the first solution are Petenwell Project on Wisconsin River and the pumped storage plant at Ludington, Michigan. The example of pile foundation is the proposed Corps of Engineer's new Dam No. 26 near Alton, Illinois on the Mississippi, where the cost of steel piles alone is estimated at 26 million dollars.

Foundation problems encountered in dam construction included grouting of pervious foundation layers; supporting a spillway on a vertical concrete arch spanning between canyon walls over a deep bed of river alluvium; and building a record 400 foot deep concrete cut-off under an earth dam. The cut-off of this imposing depth was constructed at Manicougan-3 power project in Canada. The cut-off consists of twin 2 ft.-thick parallel concrete walls spaced 10 feet apart. The walls are constructed of overlapping concrete piles, drilled and cast in place with the use of bentonite slurry. Where the cut-off is less than 170 ft. deep overlapping concrete panels were used.

The session on foundation repairs to the existing dams was highlighted by the description of emergency grouting repairs of the Old River Low Sill structure in Louisiana. A horizontal eddy that developed behind an upstream guide wall during spillway operation caused the wall to fail and this was followed by deep local scour. The scour under the concrete spillway structure in turn shortened seepage path underneath the structure and resulted in extensive "roofing" in the pile foundation.

Discussions on grouting began with a very informative talk on European grouting practices. They differ from the American practice in several ways. European specifications are much less voluminous because the contract is a performance contract. The contractor is not told how to do the job. He is expected to use his own "know-how" to produce the results. Grouting pressures used are much higher than those used in the U.S. The idea is to open rock seams, force the grout in and let the fissures close again. Grouting alluvium is more widespread than in the U.S.

The representatives of the American grouting industry present at the conference voiced their preference for the contractual approach used in Europe. It was also noted by several speakers that U.S. grouting specifications have changed little in the last 30 or more years and that they almost all conform to standard U.S. Government grouting specs which leave very little latitude for changes or innovations.

From the contractor's viewpoint good foundations are those a contractor makes money on and bad foundations are those he loses money on. This was the gist of the comments made by a speaker representing the construction industry. Naturally this assumes adequate engineering control to assure that the desired results are actually achieved. Adequate foundation information in the bid document is a must. Alternate provisions should be made in the specifications to cope with unforeseen foundation

vii

conditions. If such conditions are encountered they should be swiftly dealt with as the contractor's expenses pile up quickly even when he is only standing by. Foundation clean-up should be paid for separately.

The compact format of the conference (only a single session was held at any given time) and the intimate atmosphere of Asilomar grounds encouraged exchanges of ideas and offered ample opportunities for personal contacts.

Richard D. Harza
and
Andrew Eberhardt

INTRODUCTION

by Richard D. Harza*

Many of us in the field of dam engineering feel that the most interesting, most challenging, and potentially most dangerous part of a dam is its foundation. And often it is the most expensive part! Certainly the foundation is the part of the structure which is "least knowable in advance" and "least visible after project completion". The science-art of analysing, treating, and designing dam foundations can be compared to the practice of internal medicine. Understanding and diagnosis of the "patient" must be done with various probes and indirect tests and the resulting inferences and deductions then used to plan a course of action - the design for a dam foundation - the type of treatment or surgery for a patient. The degree of judgement required is extremely demanding in both cases, and in both cases the final proof of the diagnosis is not available until the dam foundation has been opened by excavation - or the patient has been opened by surgery.

Just as the diagnostician and the surgeon must work intimately in complementary functions, so must the geologist

*Chairman, Engineering Foundation Conference on "Foundation for Dams." Vice Pres., Harza Engrg. Co., Chicago, Ill.

and the dam foundation engineer. And that is what this conference
is all about.

This conference was the result of a recommendation made
at another Asilomar Conference in 1972 (on Rapid Construction
of Concrete Dams) that a conference on the general subject of
"Foundations for Dams" was held in 1974. The Engineering
Foundation followed this suggestion, and I accepted the
responsibility or organizing the conference. I was extremely
fortunate in persuading some of the top experts in the field
to act as Session Chairmen. The credit for the success therefore really belongs to the following persons who obtained the
outstanding presentations presented at the conference and in
these proceedings:

    Judson P. Elston   - Session Chairman for "USA versus
                                   European Grouting Practices"

    James D. Brown   - Session Chairman for "Problems of
                                   Constructing Dam Foundations"

    John Lowe, III   - Session Chairman for "Foreign
                                   Experience-Dams on Difficult
                                   Foundations"

    Andrew Eberhardt   - Session Chairman for "Foundation
                                   Maintenance, Repairs, and Special
                                   Problems"

    Dr. Ruth Terzaghi - Session Chairwoman for "Engineers
                                   and Geologists"

I wish to extend my sincere thanks and appreciation to
these persons and to each engineer and geologist who presented
a paper of a discussion at this conference. Thanks are also

due to Dr. Cole and the Administrative Staff of The Engineering Foundation as well as Mrs. Dina Phillips (my secretary) for the efficient and pleasant conference arrangements.

I believe that this conference - and these proceedings - have made a significant and positive contribution which will help us meet our obligations in the continuing need for larger, more economical, and safer dams which are frequently sited on less than ideal foundations.

CAUSES OF THE FAILURE OF THE MALPASSET DAM

Michel A. Carlier, Chief Engineer of Rural Engineering,
Ministry of Agriculture and Rural Development,
Paris, Member of the Standing Technical Committee on Dams
(France)

1 - Generalities and main characteristics of the dam

The MALPASSET dam was a thin arch-dam, curved in two planes, which was built from 1950 to 1952 in the south of FRANCE, on a small river, the "REYRAN", about 15 miles upstream the river mouth in the Mediterranean Sea (see Fig. 1,2 and3)

The height of the dam was about 200 feet above the foundations ; evacuation of overflow was assumed by a free spillway in the middle of the crest with a spillway apron downstream. A dewatering gate and a water inlet for irrigation complete the appurtenant works ; it was no hydraulic plant.

On 2nd of December 1959, during the first filling, the sudden collapse of the dam was responsible for the death of about 340 people.

2 - Inventory of the place after catastrophe

After the accident, all the part of the circular arch on the left bank, between the lowest point of the valley and the left abutment was carried away (Fig4) and the displacement of this abutment by slipping downstream was about 6 feet.

The importance of the displacements of the different parts of the dam after the accident is shown on the figure 5. The graph of the displacements of every points of the midden cantilever (Fig. 6) shows that, at the level of the foot of the dam, the displacements were more important than on the crest.

In fact, every block of the dam has turned over its superior part and the whole arch turned over its right abutment.

3 - Mecanism of the failure

The accident is now completely explained by the break of the foundations, due to uplift pressure.

If we suppose (Fig. 7) a theorical dam founded on an homogeneous rock, presenting a fault PN, the stability of the section PBN will be assured, even in taking account the uplift due to the underground water flow.

In the case of MALPASSET (Fig. 8), the foundation was formed by a gneiss with many fracturations and contorded strata in all directions. It was however, possible to discover, on the left bank, a principal fault PN which crossed the dam about 100 feet under its foundation. The single existence of this fault does not permit to explain the catastrophe.

But, immediately upstream of the dam, many little cracks were in connection and have the possibility to form a second fault in which the pressure of the reservoir have possibility to introduce in opening a continuous crack immediately upstream the dam.

In these conditions, the whole pressure of the reservoir was applied along the way BP and consequently the wolume of the rock BPN was carried downstream which was the cause of the break.

The phenomena was made easier due to the great difference of permeability of the gneiss of MALPASSET when this rock is in tension or in compression. Experimental measurements has shown that, when the rock is in compression, the DARCY's coefficient K_c was about 10^{-4} cm/s to 10^{-6} cm/s and when the rock is in tension, this coefficient K_t was about 10^{-1} to 10^{-3} cm/s.

So, for the gneiss of MALPASSET, the yield K_t/K_c is very great, between 10 and 10^5 with a medium value of 100 (for the most principal rocks, this yield is 2 to 5).

Figure 9 shows that the precedently described geological structure was existing regularly along the whole left bank.

This is the conjunction of these two faults which is the cause of the catastrophe.

The geological characteristics of the gneiss of MALPASSET and the investigations processes which were employed at the date of the construction made it impossible to discover these faults, specially the upstream's one.

The conclusion of these great catastrophe is the necessity to drain the foundations of the arch dams ; such operation was not usually done before the collapse of the dam of MALPASSET.

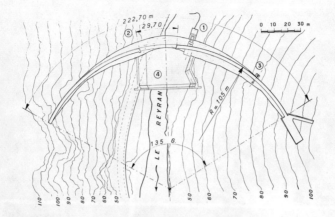

Fig. 1 : General plan of the dam of MALPASSET
(1): Outlet; (2): Spillway; (3): Water inlet; (4): Spillway apron

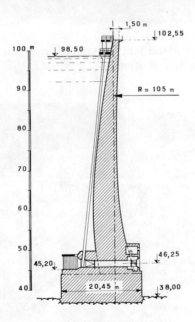

Fig. 2 : Vertical section on the dewatering gate

Fig. 3 : General view of the dam from the left river; on the first plan one
can see the left abutment.

Fig. 4 : The left bank after the failure

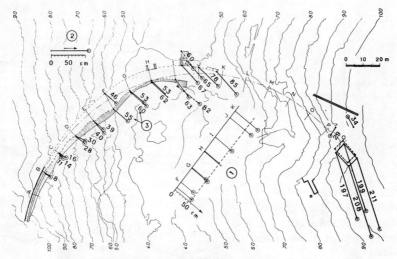

Fig. 5 : Ground plan of the displacements measured after the failure
     (1): Displacements of the inferior arch
     (2): Scale of the displacements
     (3): Displacements in cm (1 cm = 0.4 in)

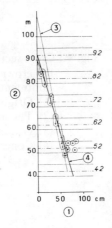

Fig. 6 : Graph of the displacements of the stations accordingly to their
     level (1): Displacements; (2): Level; (3): Joint H; (4): Average
     displacement.

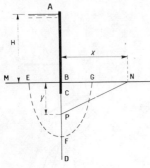

Fig. 7 : Diagram of the underflow in the supposed homogeneous rocky foundation of a theoretical dam

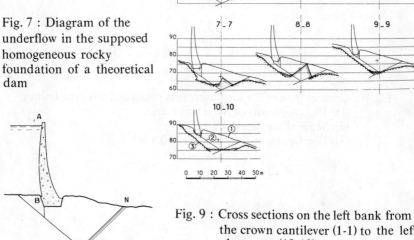

Fig. 9 : Cross sections on the left bank from the crown cantilever (1-1) to the left abutment (10-10)
(1) Original ground before failure
(2) Theoretical diedre
(3) Ground after failure
(4) Concrete

Fig. 8 : Schematic section of the dam and the foundation on the left bank

Note : The principal parts of this paper is an abstract of the book "Barrages - voûtes ; historique, accidents et incidents" (Arch-dams ; an historical, collapses and incidents) by our regretted colleague Marcel MARY, member of the Standing technical Committee on Dams. (Ed. Dunod, 1968).

by

Bruce H. Moore[1], M. ASCE

Locks and Dam No. 26 (Replacement) is located on the
Mississippi River, near Alton, Illinois, about 15 miles north
of St. Louis. (FIGURE 1)   The structure is intended to
replace an existing lock and dam, also near Alton, Illinois.
The existing locks are a 60-foot-long main structure with an
auxiliary chamber 300 feet in length. The need for replacement
of the existing structure is based upon both the condition of
the structures and prospective traffic in the river.

Lock and Dam No. 26 is located just north of the junction
of the Missouri and Mississippi Rivers. It is just south of
the mouth of the Illinois River. It thus serves as a link
connecting the Great Lakes and Minneapolis to the Ohio River
ports, New Orleans, St. Louis, and Kansas City. Traffic last
year (1972) amounted to 54 million tons. Practical lock
capacity is 41 million tons, reached in 1968. Design capacity
for the replacement is 170 million tons. Full use of this
capacity is expected before the year 2025.

Lock and Dam No. 26 was built in the 1934 to 1938 period.
Since construction started, it has been plaqued with problems.

---

1. Chief, Foundation and Materials Branch, St. Louis District,
   Corps of Engineers, St. Louis, Missouri.

VICINITY MAP

SCALE IN MILES

50   0   50   100

FIGURE 1

There were two cofferdam failures (White & Prentis
"Cofferdams"). A severe pipe developed under one of the dam
monoliths, creating voids up to 8 feet in depth beneath the
sill. The auxiliary lock wall moved in excess of 6 inches.
A scour hole deeper than the pile tips developed downstream
of the dam.

Consideration was given to repair and alteration of the
structure, but economic evaluations and engineering factors
dictated a new structure about 2 miles downstream.

The Locks and Dam No. 26 (Replacement) will have twin
separated 1200' x 110' U-frame locks (FIGURE 2). The dam will
have eight 110' x 42' tainter gates. Present plans call for
steel H piles to rock (60 long) beneath the dam (FIGURE 3)
and either H and/or displacement type piles beneath the locks
which are still under active design. Pile type for the locks
will be determined following extensive driving and load tests
during the first-stage construction contract (six bays of the
dam on the right bank). The present estimate of cost for
the structure is $380,000,000, with $26,000,000 allocated to
piles (6,000 in Dam; 20,000 in Lock).

The foundation investigations utilized a large portion
of the available spectrum of exploration tools. Existing
information was, of course, researched. Drilling techniques
included truck-mounted equipment using fixed piston undisturbed

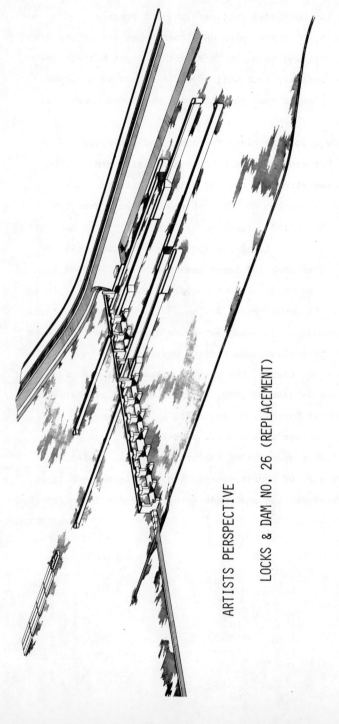

ARTISTS PERSPECTIVE

LOCKS & DAM NO. 26 (REPLACEMENT)

FIGURE 2

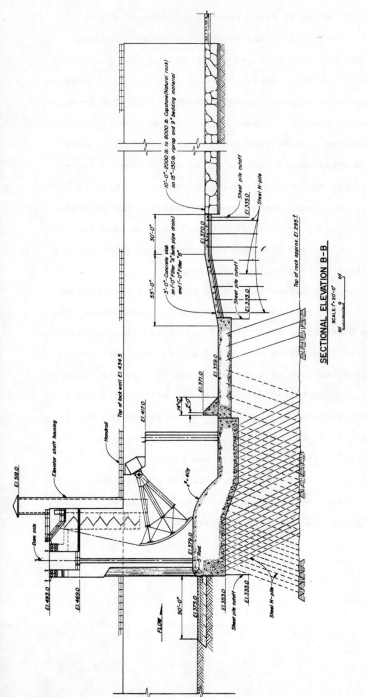

DAM SECTION - TYPICAL

SECTIONAL ELEVATION B-B

SCALE: 1" = 20'-0"

FIGURE 3

samplers, all sizes of drive spoons, and diamond coring
equipment. Churn drills and large-diameter (30") reverse
rotary equipment were employed in connection with full-scale
pumping tests on both banks. To supplement the drill explorations, overwater seismic (boomer) surveys, full-size dredging,
and down-hole electric logging techniques were used. Full-
scale overwater pile driving tests were performed. A more
detailed description of the use of down-hole logging techniques in overburden materials will follow later. This is
apparently the initial large-scale engineering use of these
logging techniques in overburden materials. In addition, the
extensive use of both standard-size spoons and 3-inch O.D.
spoons allows an interesting comparison.

A program of explorations this varied and interrelated
required a soundly established philosophy of operation. The
St. Louis District of the Corps of Engineers tries to implement
its exploration programs in logical steps. The exploration
pattern progresses in an orderly fashion through the "Survey"..
"General"..and "Detailed" design stages, with additional
information garnered at each step. At Locks and Dam No. 26
(Replacement), we have had five overwater drilling stages
supplemented by the pumping tests and seismic work. First
(in 1967) was a series of general borings which indicated that
our suppositions about the site were apparently valid. The
second stage (1970) explored the axis and found problems with

boulders. This led to the third stage (1971), investigating
the possibility of relocation. The later stages, 1972 and
1973, more precisely defined the conditions over the total
site and developed detailed information for plans and
specifications.

Initially, the exploration program was developed to
garner information directly related to the strength, density,
and pile driving resistances of an assumed pile foundation.
It was developed to provide "top of rock" information. There
was no concerted attempt to develop the "geology" of the
site, which is in the middle of the Mississippi River with
the river bluffs 1 to 10 miles away. This approach was based
upon our experience with 150 miles of levee and underseepage
investigations starting in Alton, Illinois, and extending to
the mouth of the Ohio. "River alluvium" gradually becoming
coarser with depth was the expected foundation -- like the
other 150 miles. Our concern was with the cobbles and
boulders we would encounter. The initial few borings in 1967
showed no significant deviation from the expected foundation
conditions. Three borings showed thin till above rock,
assigned as a minor remnant. Those in 1970 encountered
extreme resistance (boulders) at the upstream lock gate
chamber locations. Seismic "boomer" surveys were run at this
time to develop "top of rock" continuity. Interpretation

seemed to show three or more distinctly different zones in
the overburden. This interpretation was negated, as the then
available physical explorations apparently did not provide
support. Explorations (1971) 350 feet downstream showed
improved conditions, and a relocation decision was quickly
reached. In reviewing this exploration program, an apparent
anomoly was discovered: thin glacial till lenses within the
alluvium. The till was examined by the Illinois Geological
Survey and pronounced "Illinoian Glacial Till." A concerted
effort was initiated to explain this anomoly. All previously
collected sand and gravel samples were microscopically
examined. Minerological studies were performed. Subsequent
drilling used natural gamma, self potential, and resistivity
logging of the surficial materials. Penetration resistances
were plotted. Several differences started to appear. Grain
shape changed with depth -- rounded near the river bottom
grading to angular near the bedrock surface. Zonation was
indistinct but discernible. Carbonate content also increased
with depth. Penetration resistances showed differing zones.
Gradually, all the facts fell into a discernible geological
pattern. FIGURE 4 presents one of the finally developed
profiles crossing the river. The sands/gravels/boulders of
the lower zone when combined with the intermittent till lenses
led to a conclusion that this was one basic zone of Ice
Contact/Near Ice Illinoian deposition. The zone was

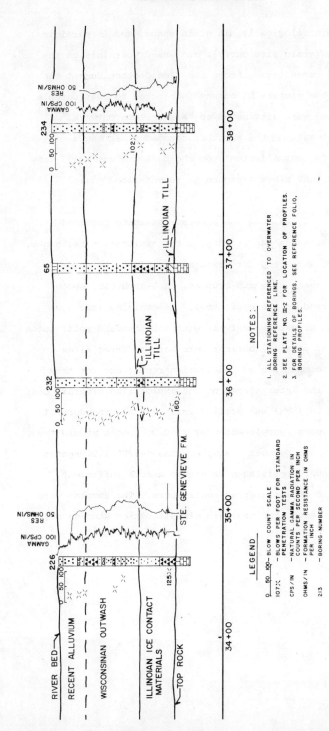

PARTIAL GEOLOGIC SECTION

FIGURE 4

NOTES:

1. ALL STATIONING REFERENCED TO OVERWATER BORING REFERENCE LINE.

2. SEE PLATE NO. III-2 FOR LOCATION OF PROFILES.

3. FOR DETAILS OF BORINGS, SEE REFERENCE FOLIO, BORING PROFILES.

LEGEND

0 50 100 — BLOW COUNT SCALE

107·X — BLOWS PER FOOT FOR STANDARD PENETRATION TESTS.

CPS/IN — NATURAL GAMMA RADIATION IN COUNTS PER SECOND PER INCH

OHMS/IN — FORMATION RESISTANCE IN OHMS PER INCH

213 — BORING NUMBER

RIVER BED

RECENT ALLUVIUM

WISCONSINAN OUTWASH

ILLINOIAN ICE CONTACT MATERIALS

TOP ROCK

STE. GENEVIEVE FM.

ILLINOIAN TILL

consistent mineralogically, in grain shape, and in particle
distribution (grain size curve), and was discernible on the
electric and gamma logs. Above the "Ice Contact Zone," the
materials fit a picture of outwash subjected to considerable
sorting. They were attributed to "Wisconsinan Outwash."
Above the outwash were the smoother rounded finer sands
assigned to "Recent Alluvium" On the bank are some real fine
sands, silts, and clays assigned as recent flood plain
deposits.

In the early exploration phases, obtaining physical
samples of the sands was primary to the exploration technique.
These samples were used for grain size analyses to guide
sampling for shear strength studies. A 3-inch O.D. spoon
sampler was used as the basic tool. Recognizing that some
people worked almost exclusively with the standard split spoon
(1-3/8" I.D.), parallel borings were made at three sites to
allow direct comparison of penetration resistance for this
project. In the later stages of the exploration program, one
program (+40 borings) was drilled using a dual procedure of
1-3/8" I.D. spoons supplemented by 3" O.D. spoons should the
smaller spoon not recover a sample. The 1-3/8" I.D. spoons
provided standard resistance values to assure sufficient
information to any interested contractors. The samples were
needed to implement the geologic classification studies being

carried forward at the same time. This was an expensive and
time-consuming drilling procedure. Subsequent programs
reverted to 3" O.D. sampler usage only.

The relationship between the standard penetration test
and the penetration test using a $2\frac{1}{2}$" I.D., 3" O.D. split
spoon sampler with 350-pound hammer and 18-inch drop (herein-
after referred to as the 3-inch penetration test) was studied.
Two groups of borings were selected for comparison. Each group
of borings was reasonably well scattered over the project area
and penetrated all zones -- the Recent Alluvium, Wisconsinan
Outwash, and Illinoian Ice Contact material. Penetration
tests for each of the three materials were grouped and plotted
versus elevation. Median lines and envelopes enclosing the
central one-half of the data points were drawn to eliminate
excessive scatter. (FIGURES 5 and 6). Although the broad
scatter of the points is attributable to many factors, such as
variability of the foundation itself, variability of field testing
conditions, i.e., hole size, condition of equipment, etc., it is
believed that the major cause of the scatter is excessively high
penetration values caused by gravel particles blocking the
spoon. This is evidenced by comparing the width of the median
50% bands in FIGURES 5 and 6.   In each case, the scatter of
the standard penetration tests was on the order of twice that
for the 3-inch penetration tests. The 3-inch penetration test

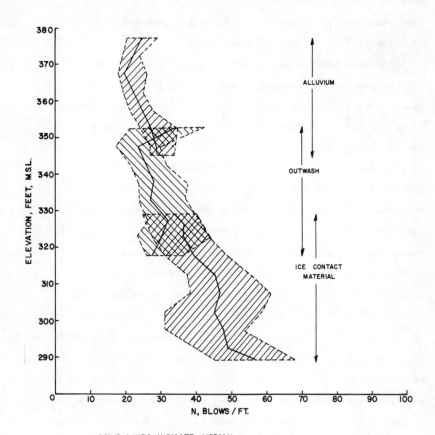

SOLID LINES INDICATE MEDIAN
N VALUES DETERMINED
5 FT. INCREMENTS

50% OF N VALUES LIE WITHIN
SHADED AREAS.

3" SPLIT SPOON
PENETRATION TESTS

FIGURE 5

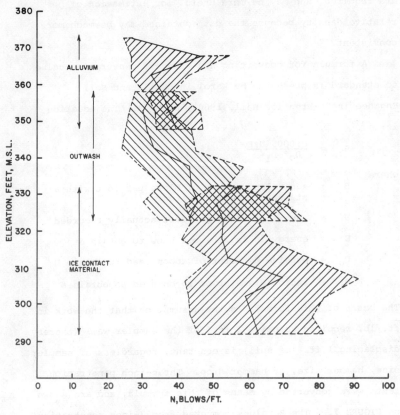

SOLID LINES INDICATE MEDIAN
N VALUES DETERMINED AT
5 FT. INCREMENTS.

50% OF N. VALUES LIE
WITHIN SHADED AREAS

STANDARD SPLIT SPOON
PENETRATION TESTS

FIGURE 6.

may be advantageous where samples of coarse grained materials
are required, and may be more useful for estimation of
relative density because the data obtained may be much more
consistent.

A formula for converting non-standard penetration values
to standard is presented by Karol in "Soils and Soil
Engineering" (Prentice-Hall, 1960), page 23. The equation
is:

$$B = \frac{0.0005 \; N \; E}{D_o^2 \; D_i^2}$$

where

B	=	equivalent number of blows per foot with a standard penetration test
N	=	number of blows per foot actually recorded
E	=	energy in in./lb. per blow to obtain N
D_o	=	outside diameter of spoon used to obtain N
D_i	=	inside diameter of spoon used to obtain N

The basis of this formula is the assumption that the work in
ft./lb. required to drive 1 ft.^{3} of the sampler wall (thereby
displacing 1 ft.^{3} of soil) is constant, regardless of sampler
size, hammer size, or hammer drop. Three-inch penetration
tests were converted by means of this formula, and are shown
on FIGURE 7. The B values computed from 3-inch penetration
tests were consistently lower than the actual standard N
values. This is apparently related to the greater ease with

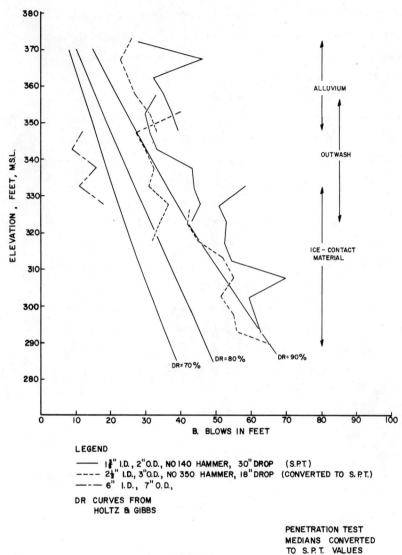

LEGEND

——— 1¾" I.D. , 2"O.D., NO 140 HAMMER, 30"DROP   (S.P.T.)
- - - - 2½" I.D., 3"O.D., NO 350 HAMMER, 18"DROP (CONVERTED TO S. P.T.)
—-— 6" I.D. , 7" O.D.,

DR CURVES FROM
   HOLTZ & GIBBS

PENETRATION TEST
MEDIANS CONVERTED
TO S. P. T. VALUES

FIGURE 7

which the 3-inch sampler can penetrate the coarse foundation material and the smaller area ratio $\left[D_o^2 - D_i^2) \; D_o^2 \right]$. In any case, use of the corrected 3-inch penetration tests (B) is slightly conservative. The curve shown in FIGURE 7 for the 6" I.D. drive sampler was based on only three borings. It is presented to indicate that the tendency for computed B value to decrease with increased sampler size, as evidenced for the 3-inch spoon, may continue with still larger spoons.

Additional research to correlate large split spoon penetration tests to relative density would be valuable in extending the Gibbs & Holtz work. In-place density tests will be taken during excavation for Locks and Dam No. 26 (Replacement) and used to refine the correlation of the 3-inch penetration test to relative density.

Electric logging was initiated into this exploration program in the hope that it would answer some questions on boulder zones. The boulders and cobbles appeared to follow the roller bit down during drilling. This created the impression that the zones were much thicker than was perhaps the actual case. It was hoped that the logger would define these zones with clarity. Such was not the case. It did, however, allow greater accuracy and confidence in our geologic interpretations over the site. Formation boundaries were discernible when samples were missed. The logger

allowed assured extrapolation between samples and throughout
some zones, where "drill action" was the defining tool with
previous techniques. With experience, formation signatures
as curve shape, baseline shift, and certain peaks were
immediately apparent. Required for effective use of the
logger was a refined mudding technique that controlled cake
thickness to less than 3/32" together with several well
sampled and studied "base" holes. The latter allowed the
signatures to be established. The logger will not provide
the classification. It will, however, provide continuity and
accuracy to the individual hole and over the formation. It
is a very valuable tool.

Pile tests performed at the original Locks and Dam No. 26
were reported by Feagin at the 1948 International Conference
on Soil Mechanics and Foundation Engineering at Rotterdam.
This was one of the first full-scale lateral load testing
programs. Pile groups were tested. Repetitive loadings were
applied. A design horizontal loading was chosen which would
not exceed 1/4" deflection. On Lock 26, the piles did not
react as predicted. The dam has apparently experienced ± 2"
of movement -- the lock, as previously noted, in excess of
6" movement, accompanied by severe cracking.

The piles at the existing dam are timber, vertical, bear
in the sands at the site, and are designed based on these

tests. The new dam will have battered steel H piles driven
to rock, or refusal slightly above rock, dependent upon
field load test results. Lock piling is still under study.

Problems facing the designer at the inception of the
project revolved around two elements: the ability to attain
sufficient penetration for assured tension and compression
loads; and the foundation soils capability to provide lateral
support to the piling. The question of lateral capacity was
an early theoretical question quickly resolved. The mathe-
matics of lateral pile load distributions is well established.
Soil resistance parameters are reasonably understood but are
not readily developed by other than full-scale testing. The
delay and attendant cost created by this need for a full-scale
test contributed to a conservative approach. Values commen-
surate with the dense to very dense soils to be found at the
site were chosen ($n_H = 100$). Erosion during construction
and other factors could greatly reduce this value. A lower
limiting case with $n_H = 10$ was used for checking design.
Repetitive loading, vibration, and group action are known to
significantly reduce these parameters. The base values of
100 and 10 were reduced by 1/3 and 1/3 again to account for
these influences. The need for extensive laboratory/
theoretical studies and preconstruction full-scale testing
was eliminated with adoption of these values.

Knowledge of the cobbles/boulders/geology at the site
was primary to successful resolution of the penetration
problem. Previous descriptions of the explorations show our
efforts in that direction. We found either boulders, cobbles,
or till all over the site. Therefore, assurance that a pile
foundation was feasible could come only from actual driving
tests. Resolution of this "feasibility" question had to come
early in the design stages. An overwater pile driving test
was chosen. The test was performed at four locations dictated
by foundation conditions. A "normal" site was first, followed
by a boulder site, an intermediate site, and finally a "till"
site. Piles (14" Steel H) were driven with three hammers
(24,500, 32,500, and 38,000 ft./lbs.). Penetration (70'-80')
was achieved at the "normal" and "intermediate" sites. The
boulders and till stopped the penetration at about 5 feet into
each zone. Various pile tips were used in driving. Extraction was attempted on all piles. The piles driven to refusal
in the till could not be extracted. The same applied to
piles driven in the intermediate zone. The reasons for lack
of extraction could have been the age and condition of the
extractor, but one contributor was the condition of the tips.
Where heavy tip reinforcement was used, there was no damage.
Almost without exception, damage occurred while attempting to
drive to refusal on the rock. FIGURE 8 shows piling after
extraction.

PILE TIP DAMAGE

FIGURE 8

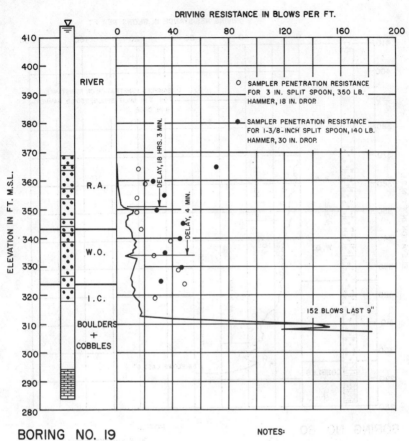

DRIVING RESISTANCE IN BLOWS PER FT.

BORING NO. 19

NOTES:
1. PILE NUMBER 2
2. DATE DRIVEN, 26 APR. 1972
3. HAMMER USED, VULCAN 010

FIGURE 9, DRIVING RESISTANCE

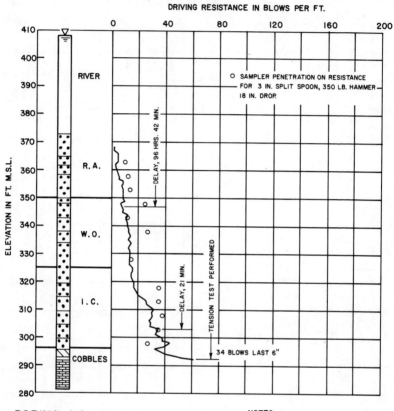

DRIVING RESISTANCE IN BLOWS PER FT.

BORING NO. 60

NOTES:

1. PILE NUMBER 5 IP-1
2. DATE DRIVEN, 10 APR. 1972
3. HAMMER USED, VULCAN 010

FIGURE 10, DRIVING RESISTANCE

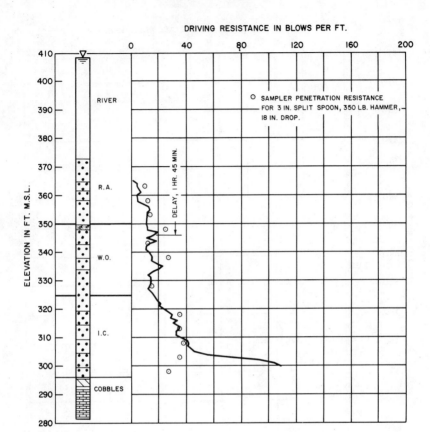

DRIVING RESISTANCE IN BLOWS PER FT.

BORING NO. 60

NOTES:
1. PILE NUMBER 1
2. DATE DRIVEN, 28 MAR. 1972
3. HAMMER USED, VULCAN 80C

FIGURE II, DRIVING RESISTANCE

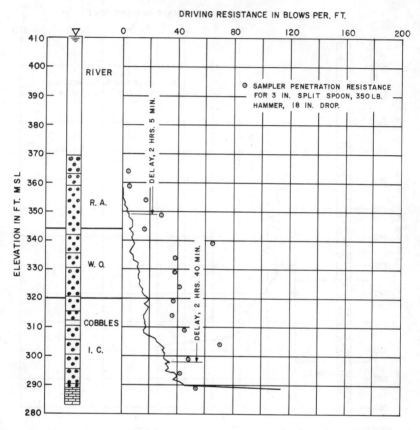

DRIVING RESISTANCE IN BLOWS PER. FT.

BORING NO. 42

NOTES:
1. PILE NUMBER 4
2. DATE DRIVEN, 22 MAY 1972
3. HAMMER USED, VULCAN OIO

FIGURE 12, DRIVING RESISTANCE

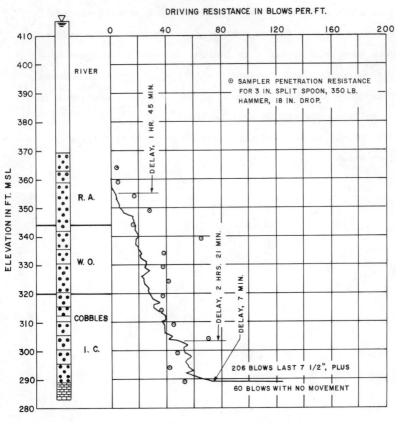

DRIVING RESISTANCE IN BLOWS PER. FT.

BORING NO. 42

NOTES:
1. PILE NUMBER 2
2. DATE DRIVEN, 12 MAY 1972
3. HAMMER USED, MKT DA55

FIGURE 13, DRIVING RESISTANCE

FIGURES 9 thru 13 are representative driving records for five H-piles driven overwater at the site. The original data have been appended to include the geologic layers of the overburden and standard (1-3/8") split spoon blowcounts. These data are for comparison of the driving resistances for 3" split spoons and HP14x73 H-piles.

An extensive load test program is planned for the first-stage construction contract.

EUROPEAN VS US GROUTING PRACTICES

Paper presented by Mr.P. Rigny

General Manager Soletanche & Rodio of Canada Ltd

The subject we want to talk about tonight is,
I know, of great interest to many of us. As the only
speaker in this session with a European background, I
will try to put the emphasis on what aspects of grouting
as done in Europe differ from the North America practice.

For this, it is necessary to deal separately
with Rock grouting and alluvial grouting.

Rock grouting

Hole spacing

No significant difference can be found here;
a typical spacing would be to start with holes 20 ft.
apart and grout them (primary holes); then drill holes
at regular intervals between the holes previously drilled
(secondary holes). As only in situ tests can show how
effective the treatment has been, this procedure may be
carried out even for them; the final spacing will be defined by the result achieved only.

Grouting procedures

As stated, a pattern of primary and secondary
holes is followed; depending on the state of the rock,
grouting is carried out in ascending or descending stages.
Whenever possible the ascending stages procedure will be
followed as it is much more economical.

Type of cement

Depending on the availibility of cement of the
right type,a firm grained cement will be chosen. It has
been found that the grain site of the cement is important
as the larger cement grain tend to form a "bridge" which
prevents further grouting of a fissure.

Washing of fissures

This is the first criterium, of those we have
mentioned up to now, where European an US practices gen-
erraly differ in a significant way. The view commonly
held in Europ is that washing of the fissures,in most cases,
can be both expensive and ineffectual. In fact as soon as

37

a communication has been established between two holes
for the purpose of washing a fissure, a "pipe" of water
is formed between those two holes and once this is done,
very little subsequent washing is achieved. So, the view
is that fissures have to be dealt with in another way.
Of course, this approach is flexible and, may differ
according to the size of the fissure, the quality of the
filling material, the purpose of the grouting, etc...

Grouting pressures

        This of course is an item where practices
differ widely and I know you are all eager to hear the
confirmation of what you know or have heard about the
use of high pressure in grouting.

        It is a fact that higher grouting pressures
are used in Europe when the 1 psi allowed per foot of
depth found in many specifications here is simply never
used.

        I have tried here to summarize some of the
reasons behind the use of higher pressure.

        - During the injection process, pressure causes
the fissures in the rock to open ,the cement particles
are deposited on the walls of the fissure during grouting
and when the pressure is realeased,there is a tendancy of
the fissure to close against thiscement there by producing a very tight bond between cement and rock;

        - Under high pressures,the grout will travel
further and this will enable to decrease the number of
holes for a given result, thus leading to a substantial
saving;

        - When a fissure is filled with material, devel-
opping pressure during grouting will compact the material
and form a lattice within this material thereby reducing
permeability and increasing overall strength of the rock
mass under treatment;

        - During grouting under pressure, water is
expelled from the grout, resulting in a higher resistance
material in place and a better bond between grout and rock.

- Rock movement Experiments have shown that even
if a higher pressure than that the computation of rock
weight would seem to allow is used, movement does not
necessarily occur in rock. In fact whether movement is
going to occur depends on the characteristics of the rock
itself and most of all on whether it has a layered
structure or not. One must keep in mind that even with
the use of low pressures, if the quantity of grout injected
is sufficient, movement will occur. This effect of creating
a float jack in the ground can be quite severe and in quite
a few instances results have been surprising and disappointing
when this phenomenon was not taken into account.

After having thus said that high pressures are
used in Europe, I think it is most important to qualify this
statement.

In fact the pressure to be used in grouting is
not defined by a theory which would tell exactly what pressure should be used, it is defined in relation to what pressure a given rock can take without fracturing or lifting,
and this can be known only from a field test.

Alluvial grouting

Alluvium grouting

For alluvium grouting I would like to emphasize
two differences between US and European practices,which I
think, are quite relevant.

The first one is the type of grouts to be used.
In Europe, it will be seen that treatment of
an alluvium calls for a sequence of grouts, each of them
achieving a certain result, and any stage of this sequence
can not be started before one is sure that those results
have been obtained. (see appendix)

For example, even if a silicate based gel is needed
to achieve a certain degree of imperviousness in the ground,
grouting of this gel will be preceded by grouting of a clay
and/or bentonite cement grout which will fill the larger
voids. The purpose of this is two fold 1.to save on the
more expansive more penetratring grout.

-2.To avoid leaching of the chemical grout in the
case of silicate basedgel or the consequences of having a
mass of grout of low resistance in the case of gels and of
some resins.

Therefore grouting an alluvium calls for a
sequence of operations where any hole can be grouted more
than once. And this is made economically possible by the

use of the "sleeved pipe" process.(see appendix)

A "sleeved pipe" is a pipe where perforation
have been made at regular spacing, say at lft. intervals,
along its length. These perforations are covered with
a sleeve. This tube is inserted in the hole, after drilling
to final depth, and then a plastic grout is injected around
it.

Grouting them takes place by using a double
packer which will enable one to inject at any desired
depth and repeat the process as often as desired in the
hole.

As I see it these two aspects a) following a
sequence from one type of grout to another b) the general
use of the sleeves pipe method,cover and explain the differences seen in US and European alluvial grouting practices.

Conclusion

Having shown a few examples of where European
and US grouting practices differ, I would like to point
out that though these differences may be very relevant to
the subject of our discussion tonight may be they are not
the most important.

As I see it, the most important difference lies
in the form of the contract to carry out the work itself.
Whereas in America a grouting contract will call for a
detailed specification of every step to be followed, in
Europe it is much more widely recognized that the procedure
has to be defined as the work goes, taking into account all
the information which is gathered during the course of this
work, and all the experience which has been gathered on
previous sites.

It is in this way that one can say grouting is
still now an art than a science.

This integration of observation and experience
calls for an involvement of the grouting contractor; whereas
if all a contractor can do is follow a set of specifications

he will obviously become passive and his management
will be management of a contract with little relevant
to the process being carried out.

To finish this talk I would like to tell you,
with the help of a few slides of a grouting job where
I think the European approach to a grouting problem,
namely observation and adaptation to the conditions dictated by the site, led to what you would not think of as
typical European characteristic : not so high pressure
and a single line of hole for a water tight curtain in
alluvium : the Cabora Bassa cofferdams.

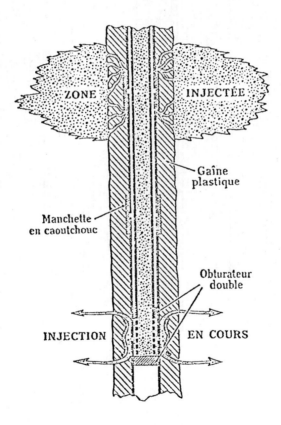

Fig. 1 The sleeved pipe process

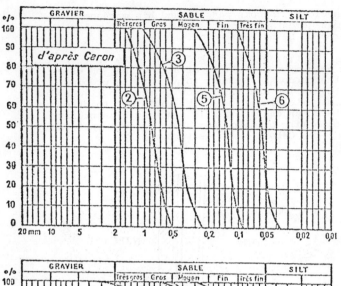

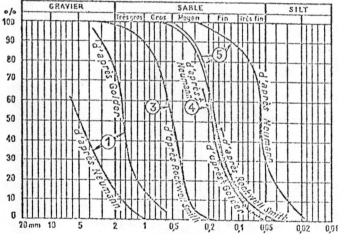

Limites de pénétrabilité des coulis basées sur la granulométrie du terrain.

1 — ciment;   4 — procédé JOOSTEN;
2 — argile-ciment;   5 — gel de silice;
3 — argile;   6 — émulsion de bitume-résine.

Fig. 2 Limits of injectability:
        Types of grout and
        ground penetration

Fig. 3 General view

Fig. 4 More detailed view of ground to be grouted.
Alluvium and blocs to be grouted in the
river bed were a maximum 50 ft. thick.

Fig. 5 View of rock abutment

Fig. 6 Partial view of grouting station

Fig. 7 Water flow that had to be stopped

Comments on U.S. Grouting Practices

James M. Polatty, P.E.
Retired: Chief Engineer, Mech. Div.
U.S. Army Engineers
Waterways Experiment Station
Presently: Consulting Engineer

This session or United States versus European Grouting practices
should be most informative for everyone engaged in the grouting profession. I firmly believe that there has existed for years an excellent
exchange of information between grouting people in America and their
counterparts in both Europe and other countries of the world. This is
evidenced by the papers that have been presented or published by the
ASCE and AEG. A study of European literature indicates that personnel
from North America are presenting papers at European meetings concerned
with grouting; therefore, there is a partial exchange of information.

Mr. H. Fabregue adequately described the difference in American and
European practice at the ASCE meeting at Mobile in 1965 when he stated
for large-scale grouting the difference is contractural. In the U.S.
the preliminary investigation and design leading to the grouting are
generally carried out by the owner who prepares detailed specifications
and advertises for contractors to bid. In Europe, usually the contractor
carries out the entire program and guarantees the work. It would seem
that the typical American procedure describes the Bureau and Corp grouting,
while in America perhaps the TVA with its hired labor force and Halliburton
with its packaged job may approach the European practice. Of course there

are, from my viewpoint, advantages and disadvantages to both approaches.
Indefinite costs is usually the disadvantage offered to the so-called
European method, while the American system is limited in what the contractor can do to obtain a better job or save money. One great
disadvantage to the European grouting people have been their tendency to
keep the details of their jobs rather secret. This is, of course,
understandable as this is what their company is selling, but it isn't
a good practice if we desire to increase the knowledge of the entire
grouting profession and secure better grouting jobs.

I have been working in the field of dam constructing for the last
35 years and have, in this time, been exposed to a number of grouting
jobs covering a wide range of work. Our ASCE committee decided some
years ago that grouting should be broken down into a number of special-
ities that require either special equipment or material and experience
in performing this special task and that work was needed either by
research or a literature study in the following:

a. Professionalism in grouting - to increase the over-all
professional knowledge of personnel planning and performing the
grouting operation.

b. Bibliographies - to furnish a source of literature to people
planning grouting jobs.

c. A review of European grouting practice by translations and
report.

d. Grouts that will not shrink or harden and ones that will have a controlled expansion.

e. Procedure for grouting abandoned mines.

f. Grouting preplaced aggragate to form concrete.

g. Pressure grouting of highways to restore their load carrying
ability.

h. Grouting for slab jacking and stablization.

i. Grouting of solution cavities or channels.

j. Repair of concrete by grouting.

k. Grouting allurial material.

l. Grouting form.

m. Sponsor sessions on grouting.

n. Co-ordinate with other professional organizations having an
interest in grouting.

o. What are the problem areas where grouting may offer a solution.

p. Evaluate the feasibility and the effectiveness of grouting.

In preparing this paper, I requested copies of current specifications for foundation grouting from several Corps of Engineers districts,
the TVA and the Bureau of Reclamation. In comparing these current
specifications with copies of specifications that I had in my file
that are 30 years old plus my observations and experience, I concluded
that we in the United States have not, in general, changed any of our
approaches on grouting. And, this is good. We have a large number
of highly successful jobs that have performed satisfactorily over the
years; therefore, the design and procedures have been well proven.

It would appear that Corps of Engineers grouting can be roughly divided into the following:

a. Grouting for water control during construction.

b. Curtain grouting.

c. Area or blanket (remedial foundations).

d. Repair work.

Well, what's new?

The Department of Defense has decided that its organization will include in its specifications a Contractor's Quality Control program along with a Government quality assurance program. The attempt is for the Government to stop directing how the contractor does his job in complying with the specifications requirements. The contractor, by the job specification, is required to set up a quality control organization with the Government being responsible in assuring that the required quality is being obtained.

This program, based on reports from our field people, is both working and not working. In general, good results are based on the intent and desires of the contractor's organization, but still the responsibility or assurance for a good job rests with the Government.

Value engineering is another attempt by the Corps to reduce costs. By this contractural opportunity the contractor can recommend an alternate procedure, technique or materials that can serve as a substitute for those required by the specifications. The cost savings are split between the government and the contractor.

Curtain grouting

For one of our Corps of Engineers job awarded this calendar year, the specifications call for a multiple line grout curtain. As is customary, the grout plant's location is specified as to maximum allowable distance from the hole; the plant must deliver, supply, mix, and pump a mixture to the satisfaction of the Contracting Officer. The capacity of the plant is specified along with line size and pressure, gage precision (high), plus a requirement for a gage testing capacity by the contractor.

Materials include the usual sand, cement, and water, all complying with specific requirements, except that the cement _must_ be supplied in bags. The sand must meet a rather tight grading specification.

Each row of the curtain is to be developed by the split spacing, zone, and stage method which is specified in detail. Drilling in some selected area is required to be accomplished with rotary diamond bits with percussion being allowed in certain areas.

The specification is implicit as to the embedded headers, the definition of terms, the procedures for washing, pressure testing, and stage grouting, and the requirements for packers.

One hooker is caulking, a nonpay item, that could possibly vary considerably in cost, depending on the approach of the inspectors. It is my opinion that the owner can obtain far better prices if, in his specifications, he will list as many pay items as possible and minimize grouping items of work.

Water control during construction

Most contracts for concrete work and all for soils usually require
that the construction work be performed in the dry. In areas of high
water table it is necessary to construct a cofferdam; the usual
procedure is to grout below the cofferdam to reduce pumping of water
and to insure stability of the cofferdam. The decision as to the amount
of grouting to be performed so as to minimize water pumping cost
requires considerable engineering judgment or a good crystal ball.

Area or blanket grouting (remedial)

This type of grouting has been used for all types of structures,
including dams and buildings. I suppose we can say we have used up
most of the good foundation sites, with remedial treatment being
necessary for a large number of structures. One problem area is in
cavernous limestone. The TVA, one of the first to encounter this
problem, uses mining and air-water washing to some degree of success.
The Corps has had problems in this area also, and to my knowledge,
no positive solution exists today, with the treatment for each
particular site being selected by the engineers and geologists
representing the owner.

Compaction grouting

The use of grouting procedures to increase the density of a soil
mass has been used for some time. A refinement of the technique of
using a growing bulb of grout to move soil particles into a closer
spacing has been accomplished recently and appears to offer considerable

advantages in special conditions.

Non-shrink grouts

Foundations for heavy machinery and special buildings have required grout treatment to insure their stability. The grout mixture must not shrink in volume and must have adequate strength characteristics.

Repair work

The use of grouting for repair work has shown a considerable increase in the last 10 years. The repairs may be required for a number of different reasons, but have been successfully used to restore the structure so that it can safely perform the task for which it was designed.

Chemicals

There are a large number of commercially available materials on the market today that are used in grouting. The Corps of Engineers, in order to increase the knowledge of the material to its engineers, has published an Engineering Manual on Chemical Grouting.

Most of the materials combinations classed as chemical grouts are relatively expensive when compared with portland-cement grout, but they have a number of extremely useful characteristics, such as:

1. Some will penetrate small voids or fissures that portland cement cannot.

2. Some will penetrate small voids or fissures that only water can penetrate.

3. The time of setting can be controlled by varying the percentage

of component chemicals.

4. Strength and durability may vary considerably, depending on the chemicals and systems.

Problems

1. Communications and training:

    a. Between the designer and the field.

    b. Between the contractor and owner.

    c. By the contractor quality control personnel and his grouting personnel.

    d. By the field personnel - there is a shortage of experienced people.

    e. Between the contractor and his sub-contractors.

    f. Inspection of drilling operation in order to obtain useful information as to characteristics of the foundation.

    g. Accuracy of pressure gages used in grouting.

2. There is some difficulty in having sufficient flexibility in the field to make necessary changes to insure a good grouting job.

3. Water-cement ratio of the portland cement mixture has to be kept low.

4. Complete records are extremely necessary.

5. Experienced and inexperienced personnel.

    a. Contractor and resident offices.

    b. Owners personnel not properly trained.

    c. On some jobs, experienced personnel are required by the

specifications.

6. Getting people to talk about their problems.

At the Waterway Experiment Station we have exchanged contracts with libraries all over the world, making available a wide selection of grouting literature. The loan of these publications are available to the engineering profession by merely writing to WES.

GROUTING: THEN, NOW AND WHERE

J. F. Daly

Resident Engineer

Bechtel Associates

When I received the invitation to write a paper to be presented at this conference, the first thing that came into my
mind was what should be my subject. I have not, in the last
four and a half years, been directly involved in a project that
required any foundation grouting. When I realized that, my first
inclination was to prepare a paper based on past experience with
a particular project. I did some research through papers and
reports that I had previously prepared and also papers that have
been written by others on foundation grouting, grouting for dams
and grouting for tunneling. Not exactly to my surprise, I discovered that the great majority of the published papers were
prepared based on a given project, or several given projects, that
the author or the author's company had participated in. There
is, of course, the second type of paper that I ran across, that is,
the highly technical paper based on theories, laboratory work,
etc., during developmental periods for the various proprietary
chemical grouts.

After considering this, I concluded that, because I am not
in a position to prepare a paper based on a recent grout job of
a recent project, nor do I have any proprietary processes or
grout formulas to discuss in a technical paper, that I would prepare a synopsis of the history of grouting in the private sector
of the construction industry, as I know it, from the time that
I was involved with my first grout job.

Fortunately or unfortunately, as the case may be, my first experience with grouting was in the late 1940's as a drillers helper for a drilling and grouting company that was under contract with a midwestern municipal water supply district. This stretches back a long way and as a drill helper I did not have all the access in the world to all the facts of the project, but as I remember it the particular municipal water resources department either had aid in the form of inspectors from the Corps of Engineers, or were using Corps of Engineers prepared specifications.

The drilling and grouting, from my standpoint at that particular period of time, was quite simple. The specifications under which we worked told us where to drill the hole, what size that hole should be and what depth that hole should go to. It then told us that we should inject a neat cement grout of a certain mix into the hole under certain pressures until such time as a given amount of cement solids per given period of time was being injected.

The specifications for this particular project apparently told the contractor exactly what type of drill to use, without reference to trade names, the type of pump, the size of hose, the size of fittings, and the arrangement of fittings on the grout tree. All of these factors were to be followed exactly. The grouting contractor for this particular project was in fact no more than a labor-equipment broker. As long as he followed the guidelines as stated in the specifications he had no problems.

In the late 1950's, I had become, through attending college, an engineer employed by Bechtel Corporation of San Francisco, Calif. Because I had worked as a laborer on drilling and grouting previously, it apparently seemed quite natural that I should be given assignments dealing with exploratory drilling and engineering evaluation of hydroelectric sites based on the exploratory work. Later when the construction of these sites commenced, I was, in general, charged with the foundation treatment program in the field.

The first set of specifications that I did get involved with, in early 1958, were quite similar to the specifications that I was introduced to in the late 1940's. I discovered during this period that to properly accomplish a drilling and grout program for a dam, foundation of a powerhouse or other structure, there were several things needed besides a rigid blueprint of equipment, mixes, pressures, and hole completion criteria.

When I was asked to prepare, in the field, a grout procedure to be followed by a contractor for an earth fill dam, I found that to accurately and adequately prepare this document I must know the purpose of the grouting. Is it for foundation seepage cutoff? Is the grouting to be used as a shallow blanket grout? Is it to be used as consolidation grouting? What is the purpose? I then found that to be able to intelligently pursue this grouting procedure I would also have to know what were the designers basic assumptions. Inasmuch as, an old professor of mine told me at one time, all dams leak, all foundations have the capability of seepage.

Therefore, two criteria must be satisfied for the design and construction of a successful hydroelectric complex. The basic criteria, of course, is structural integrity, the assurance that a given foundation is capable or can be made capable of safely supporting the structures involved. The second criteria to be considered is economics. Assuming that the structural integrity is not affected by seepage around, under or through the structures, there must be a limit as to the amount of water that can be allowed to escape or seep away without affecting the economics of the project.

After I had researched the many elements involved in the preparation of an effective grouting procedure for a given site, I would then prepare the document which, hopefully, would satisfy the design criteria and also be workable. As I look back I found that of the more than several grouting procedures that I did write, they were as varied in content as the physical characteristics of the sites that they were applied to. But also as I look back, I find that the specifications that had been produced by the engineers were for all practical purposes identical to the first set that I had worked with in the 1940's. In the mid-1960's, we did modernize our thinking enough to permit the use of pumps for pressure grouting other than a duplex piston air driven pump. This was quite a milestone from the construction engineer's standpoint, for it recognized the fact that there existed foundations that were structurally acceptable, which were not necessarily sound hard rock, but could not tolerate the pressure surges inherent with the use of the duplex piston pump.

The type of specifications and procedures described above
apparently produced structural foundations that satisfied the
designer's criteria. But the fact must be taken into consideration, that on all the sites that I was involved with, the foundation materials were generally sound competent rock. The challenge
was in shutting off springs and decreasing potential seepage paths,
not in constructing an adequate structural foundation.

The title of this paper is "Grouting: Then, Now and Where."
I have just presented my reflections on the "Then" method of
writing specifications for foundations grouting and putting these
specifications into practice on a given project at a given site.
The "Now" aspect of my paper comes, not so much from personal
involvement, but more by reviewing some ten to fifteen different
engineering firms' recent grouting specifications.

A review of several specifications for the drilling and
grouting of dam foundations, written within the last four years,
revealed that basically there is little change in concept between
these specifications and those written in the 1950's. The curtain,
or cutoff, grouting procedures are spelled out in the specifications, with respect to depth of hole, size of hole and spacing of
holes. The details of the actual grouting program are not
completed until excavation has progressed, along with some drilling
and grouting being accomplished. The innovation of the "Now"
specifications is within the equipment field where we apparently
have become sophisticated enough to allow various types of drills
and various types of pumps to be brought on the job as long as they

will perform their function in the manner that will achieve a
successful grout program.

Having been recently assigned to the Washington, D.C. area,
to serve as a Resident Engineer during the construction of the
Washington Metropolitan Area Transit facilities, I have had the
opportunity, on one project, to become familiar with specifications
written by several engineering design firms. In reviewing these
specifications for tunneling and foundations of structures other
than dams and powerhouses, I have found a very high degree of
reliance being placed upon the final result type of specification.

The contractor is being told for what purpose the grouting
is intended and in some cases the contractor is being told in the
specifications what the final or desired shear strength of a
material has to be to satisfy the specifications. He is then
being given the task, by specification, of preparing the actual
procedure that he will follow in order to accomplish these goals.
These procedures are then subject to review and approval by the
design engineers. Although the specifications do not generally
contain all of the preliminary investigation reports, tests, and
design data leading to the requirements of the final grout program,
this data is made available to the contractor at any time during
the bidding period.

The results of producing this type of specification has so far
been variable. Some contractors have elected to prepare, within
their own shop, the grouting procedure and equipment lists. These
contractors, in general, will do their necessary grouting with

their own forces and equipment. Other contractors have elected to
subcontract this complete specification requirement to a specialist grouting subcontractor, whereby the specialist subcontractor
working with the information available to all contractors will
prepare a grouting procedure and actually bring their equipment
and their technical personnel to the jobsite to perform the necessary grouting activities.

It is recognized that basic differences exist between the
grouting for structural foundations and that done for tunneling.
It has been pointed out that the foundation treatment performed
for a hydroelectric project becomes a permanent part of the structures, while the grouting accomplished for tunneling and open-
cut excavation is actually a contractor's tool used to aid in the
construction effort.

This point, being made, infers that a great deal of technical
design engineering is expended on the specification requirements
for the grouting of structural foundations, while this is not
necessarily the case for the construction expedient type of grouting.

The final portion of this presentation is the "Where." This
can be broken in two parts. For the first part, it must be asked,
where are we now in the arts of grouting from an engineering standpoint. It has been long known that the sites best suited for
hydroelectric development were fast being used up. There has been
some comment in the last few years that we will be forced to
utilize sites with foundation conditions which are somewhat less

than ideal if we are to continue to expand hydroelectric production.

Because of this, it would seem that the engineering designers
are faced with the problem of soul searching with respect to the
grouting of a foundation for a dam or powerhouse. Can we satisfy
ourselves that by using the techniques that we have learned in
the past, in the United States and also by reaching across the
ocean for materials and techniques that have been used over there
for sometime, that we are prepared to design a restructured
foundation for a hydroelectric complex? It would seem that, with
as much experience that has been gained over the years not only
in the construction of dams and powerhouses, but of other major
construction activities where varying classifications of soil
materials have been successfully treated by grouting methods,
these techniques can now be safely utilized for founding dams and
powerhouses when the proper considerations are given to the design
of a grouted structure to support these dams and powerhouses.

If we are not in a position to confidently design a restructured foundation to meet our needs, then it would seem that
the last part of the "Where" question, which is simply, "where
do we go from here," may be answered just as simply in one word:
nowhere.

PRESENT STATE OF THE GROUTING INDUSTRY

IN THE UNITED STATES

by

Joseph P. Welsh[I]

This Conference has shown that the designers of U. S. dam
foundations are not taking complete advantage of the latest technology available in the grouting industry. James Polatty, chairman
of the ASCE Committee on Grouting, indicated that a survey he conducted showed very few changes in Federal Specifications on grouting
in the past twenty years. James Daly of Bechtel notes that he has
seen a change only in the type of grout pump. Marcel Rigny of Solrod
discussed European grouting techniques and alluded to the fact that
the Europeans are utilizing much more advanced grouting techniques
to solve dam foundation and other soil and rock problems. It is my
feeling that these and other techniques and materials are available
in the United States but are not being adequately employed. I'm not
suggesting that time-honored and proven techniques of grouting be
chucked, but rather that the latest techniques, materials and methods
be investigated for economy and feasibility for future projects where
grouting may be a solution.

From a volume standpoint, a majority of grouting operations in
the United States are performed for government agencies who utilize
a competitive bid basis to select the grouting contractor. In effect,
these contractors are suppliers of labor and material who are instructed
in all procedures to utilize from the project specifications and government personnel; e.g., mix design, pumping pressures, rate of injection,
etc. Each grouting operation is modeled after previous projects and
innovations are frowned upon.

I - Eastern Manager, SOILTECH Dept., Raymond International Inc.

Another type of grouting contractor has evolved in the construction industry - principally to service private industry and
non-governmental agencies. These specialized grouting firms have
thrived based on their ability to solve unusual foundation problems by new methods and techniques. Space will not allow complete
discussion of all the latest developments of these specialized
firms, but I hope the following will whet your appetite:

(a) A method of drilling whereby the percussion drill
     and the grout tubes are installed together. The
     drill steel is extracted, a double packer isolates
     over the pre-set valves in the casing, and grouting
     is accomplished. This technique has been used in
     the United States to grout formations where conven-
     tional drilling would be very difficult.

(b) A method of installing grout pipes through dense over-
     burden ($N > 100$) at the rate of one foot/second (0.3048 m/s).

(c) Creating an impermeable cutoff wall utilizing a combi-
     nation of cement/chemical grouts. One such wall was
     created in the drawdown of an active deep well, after
     which saturation chemical grouting was accomplished
     adjacent to the cutoff wall to a depth of 100 feet
     (30.48 m) and a previously installed adjacent drainage
     blanket was not contaminated.

(d) New grouting techniques to fill voids include use of chem-
     ical/cement grouts to instantaneously set, even in flowing
     water. In effect, the chemical grout gels upon contact
     with the cement and holds the cement in suspension until
     hydration takes place - an ideal way to seal off flowing
     water beneath existing dams.

(e) Use of predesigned fabric containment vessels which
     are filled with grout or concrete to enable grout
     or concrete to be placed under water without dilu-
     tion by water.

(f) Use of fabric forms inflated with grout to line dams,
     dikes, canals, etc., both above and below water, to
     prevent erosion.

(g) Concerning chemical grouting, Joosten began this in-
     dustry utilizing his two-shot silicate process in
     1926, and in the past fifty years chemical grouts
     have become available which will penetrate into any
     formation where water may travel, will set at a pre-
     determined time and will produce strengths in grouted
     soil over 1000 psi (6.9 kPa). Chemical grouts are
     still more expensive than cement, but with rising
     labor costs the cost differential is decreasing.
     The Corps of Engineers' manual on Chemical Grouting,[1]
     which James Polatty mentioned, adds a degree of recog-
     nition to this technology but consists of nothing more
     than a summary of existing chemical grout data.

With the present limited market for chemical grouts and the
reluctance of government agencies to promulgate and research these
materials, the chemical industry will expend no additional monies
for development of chemical grouts unless by chance one is developed
for another use. Therefore, new chemical grouts are being utilized

Appendix I - References
(1) "Chemical Grouting" - Office of The Chief of Engineers
     Engineering Manual 1110-2-3504, 31 May 1974

which have not been completely tested and evaluated. Future
innovations will be slowly advanced by the grouting industry at
the expense of private industry.

CONCLUSIONS

New grouting techniques and materials are available in the
United States but are not being utilized by governmental agencies.
Further, although private industry is advancing grouting technology,
unless governmental agencies actively contribute to the research
and development in this industry and utilize the latest technology,
efforts to advance this art to a science will be slow.

DAM FOUNDATION PROBLEMS

AND

SOLUTIONS

B. I. Maduke

Harza Engineering Company

INTRODUCTION

Three examples of water resources projects completed by the
Harza Engineering Company during the 1940's and 1950's are described
herein. They are early examples of projects where other than the
most desirable foundation conditions were found to exist.

The nature of the foundation conditions are described in general.
The design solutions developed; the construction procedures employed
and the performance of the structures after completion are discussed.

PROJECT A.

Project A is a hydroelectric project located in the south eastern part of the State of South Carolina on Broughton Hall Creek which discharges into the Cooper River north of Charleston. The project was completed in 1942. The principal structures consist of a navigation lock with a 75 foot lift; a powerhouse with an installed capacity of 130,000 kw and several miles of embankment dams and dikes on either side of the concrete structures. The general arrangement of the structures is shown on Exhibit A at the back of this paper.

The dams and dikes were homogeneous sections constructed of a sandy clay and were founded on about 20 feet of clays and sands overlying marl (calcareous clay) and soft limestone. An upstream cutoff trench penetrated into the marl formation. The powerhouse and lock were founded on the soft limestone.

Investigations prior to construction, consisting of many borings and surface mapping of visible sinkholes, revealed that cavernous conditions, within the overburden immediately above the limestone, existed along several miles of the embankment dams and particularly along the east dam alignment and that some method of overcoming this hazard would be required. During the early part of the construction of the earth dams this hazard was dramatized when a 300 square foot area of ground suddenly dropped some 20 feet below the existing ground surface.

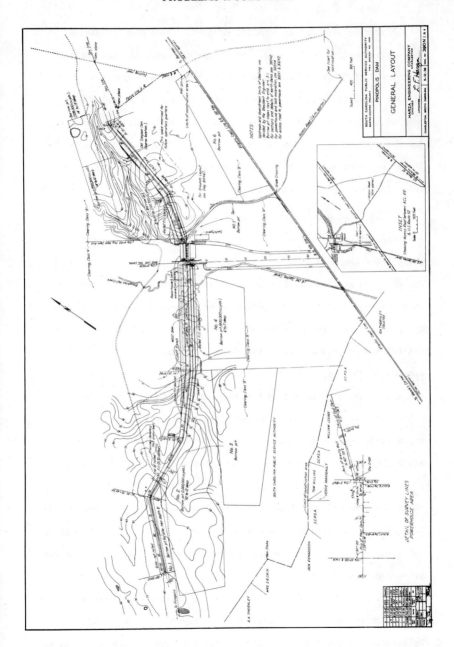

A thorough foundation grouting program to seal the foundation was carried out along the alignment of the dams adapting the earlier experience of others at the Madden dam in the Panama canal zone where similar foundation conditions were found to exist. Clay grout was used at the Madden dam to seal the foundation cavities. The grout for this project consisted of a closely controlled sandy clay with 10 percent cement, by volume, added to the grout mixture. The following grouting procedure was employed.

Areas to be grouted were selected after careful study of the locations of known sinkholes and the results of borings. In those areas where the head of water was greater than five feet and the need for grouting was indicated, primary grout holes were drilled and grouted at a spacing of 80 feet. Secondary holes were drilled mid-way between the primary holes reducing the spacing to 40 feet. Tertiary holes were drilled and grouted mid-way between the secondary holes decreasing the spacing to 20 feet. Twenty feet was considered the maximum final spacing in any area providing the grout holes took less than five cubic yards of grout. Where the grout take was greater than five cubic yards grout holes were spaced at 10 foot centers. A minimum spacing of five feet between holes was resorted to when the holes at 10 foot centers continued to take more than five cubic yards per hole. All holes were drilled 25 feet into the limestone; all previous borings indicating that there were no cavities below this elevation. The grout holes were drilled along the upstream toe of the dams as the latter were being constructed. In areas where exceptionally large grout takes were recorded, additional holes were drilled and grouted along the centerline of the dam to intercept any caverns which may not have been fully filled by grouting from the upstream toe. A total of 1,800 holes were drilled and grouted with 21,000 cu. yds. of grout being used. The subsurface grouting was supplemented by blanketing surface sinkholes, upstream from the toes of the dams, with impervious clay soils.

Upon filling of the reservoir no significant foundation seepage developed downstream of the dams and dikes indicating that the grouting and blanketing program was essentially a success. Recent reports from the site indicate that though foundation seepage still exists it does not exceed that which occurred upon the first filling of the reservoir.

PROJECT B.

Project B is a low head hydroelectric project completed in 1952 and located on the Wisconsin River in the State of Wisconsin. The principal structures consist of a 140 foot long powerhouse; a 550 foot long spillway structure and about nine miles of embankment dams and dikes. The maximum head is about 45 feet. The general arrangement of this project is shown in Figure 1.

The foundation conditions for the entire development consist of fine to coarse alluvial sands having a maximum thickness of over 120 feet. Laboratory tests gave values of the coefficient of permeability, varying between 3×10^{-2} cm/sec and 2×10^{-1} cm/sec at void ratios of 0.46 and 0.60 respectively. The natural dry density of the sands, to the depths explored, varied between 95 pcf. and 112

Fig. 1-Plan of Project Site

pcf. Because of the great depth to bedrock all structures were designed to be carried directly upon the sand. Because impervious borrow materials were not available within an economical haul distance of the site the dam and dikes were designed as homogeneous sections constructed of the readily available alluvial sands.

Since the structures were to be founded on a fairly competent but pervious foundation material the following design problems required solution; a) safe foundation loading; b) control of settlements; c) control of uplift pressures; d) control of seepage flows; e) safety against sliding; and f) control of erosion downstream of the spillway structure. These design requirements were accomplished in the following ways.

The powerhouse and spillway structures are founded on a monolithic continuously reinforced concrete slab about four feet thick bearing on the sand foundation. The foundation stresses due to the concrete structures did not exceed 3000 psf. The settlement of the structures was estimated to be in the order of 1/4 inch per 1000 psf. of applied load, the majority, if not all of which would take place during construction. The structural continuity of the foundation mat would limit differential settlement along the structures to acceptable values.

The control of uplift pressures and seepage flows beneath the concrete structures was accomplished by the provision of an upstream reinforced concrete apron; partially penetrating steel sheet piling at the upstream end of the apron and a positive drainage system downstream of the longitudinal centerline of the structures. These features are shown in Figures 2 and 3. The flow net shown is for conditions assumed during the design stages. The efficacy of the upstream concrete apron in reducing uplift pressures and seepage flows was dependent upon the imperviousness of the construction joint

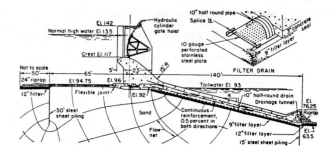

Fig. 2-Cross Section of Spillway

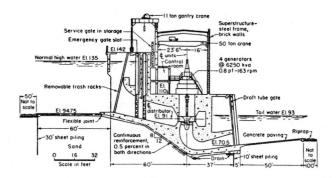

Fig. 3-Cross Section of Powerhouse

between the downstream end of the apron and the structures. A
structurally continuous, flexible and impervious joint was provided
which permitted the frictional sliding resistance of the concrete
apron to be active in the completed structure. Drainage beneath
the downstream apron of the spillway is controlled by a continuous
layer of filter material and drains spaced at 17.5 foot centers
discharging into a drainage tunnel at the toe of the apron.

The dam and dikes were homogeneous sections constructed with
alluvial sands in lifts of 12 inches which were compacted by the
traffic of the construction equipment. A filter drain along the
downstream toe kept the phreatic surface within the cross-section
of the dam, see Figure 4.

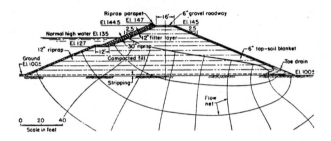

Fig. 4-Cross Section of Main Dam

Erosion downstream of the spillway structure was controlled by the
provision of a sloping apron and deflecting sill at the end of the
apron. Steel sheet piling along the downstream toe of the spillway
apron guarded against the loss of foundation support in this area.
See Figure 2. This design evolved from an extensive series of
hydraulic model tests.

Since completion of the project no measurable settlement or
differential movements have been recorded.

Typical uplift pressure conditions for various reservoir levels
under the powerhouse and spillway structures are shown in Figures 5
and 6 respectively.

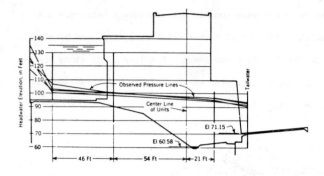

Fig. 5-Pressure Levels In Powerhouse Section

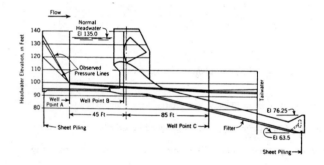

Fig. 6-Pressure Levels In Spillway Section

These readings show that the flexible upstream joint between
the upstream apron and the structures is in fact impervious. The
significant drop in uplift pressure between the upstream end of the
apron and the first observation point is greater than that based on
design analysis which assumed uniform permeability in all directions.
These readings indicate that the vertical permeability of the foundation sand is less than its horizontal permeability.

The phreatic surface through the embankment was measured by means of observation wells at several sections along the downstream slope.   The results at two such sections are shown in Figure 7.

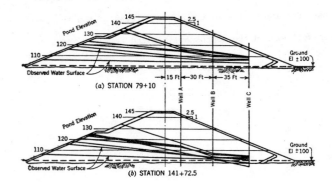

Fig. 7-Pressure Levels in Main Dam Section

In general the phreatic surface was found to be in good agreement with that found from the flow net analyses prior to construction. Leakage through the dam and foundation could not be measured but based on the measured phreatic surface it would be about that which was expected.   Some boils did develop downstream of the toe of the dam because of the greater permeability of the foundation materials and these were blanketed with filter materials to prevent erosion due to piping.

F.P.C. inspections of the project in 1967 and 1972 indicated that the seepage through the foundations of the dam and dikes had decreased to the point where none is now visible.   It is believed that this decrease in seepage is due primarily to deposition of fine grained sediments in the reservoir. The uplift pressure conditions under the spillway and powerhouse have stabilized at values equal to or less than those measured when the project went into operation. The stability of all structures has therefore, improved over the years. Erosion of the river bed downstream of the spillway apron has been well within the depths anticipated from the model tests.

PROJECT C.

Project C is a run-of-the-river low head hydroelectric project
located in the State of Washington on the Pend Oreille River. The
principal structures consist of a main concrete spillway-dam and
powerhouse with a maximum gross head of about 45 feet and an installed
capacity of 60,000 kw. The construction was completed in 1955. The
general layout of the project is shown in Figure 8.

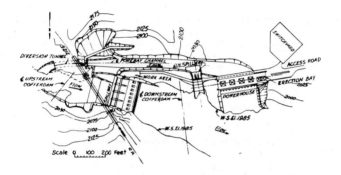

Fig. 8-General Plan

The project is located at the downstream end of a narrow canyon
with nearly vertical rock walls. At river level the canyon is about
170 feet wide. The rock walls consist of a competent marbelized and
dolomitic limestone. At the location of the spillway dam the depth
to bedrock in the stream channel was in excess of 150 feet below the
normal river level. This channel was infilled with compact to dense
fine to coarse sands and gravels with a 10 foot thick horizon of sands
and boulders forming the surface of the river bed.

The forebay channel, auxiliary spillway structure and the
powerhouse pass through and are founded on the limestone bedrock
along the left bank of the river. The spillway dam is located
over the buried river channel and required special consideration.

Various design and construction procedures were studied and
these led to the development of a unique foundation for the spillway structure which eliminated the settlement problem, guarded against
piping and erosion and reduced the uplift pressures along the base
of the structure to safe values. The spillway structure was supported
on a reinforced concrete arch spanning across the compressible and
pervious alluvium of the river bed. See Figures 9 and 10.

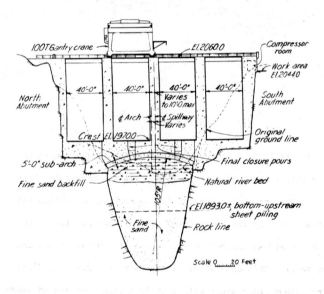

Fig. 9-Main Spillway-Longitudinal Section

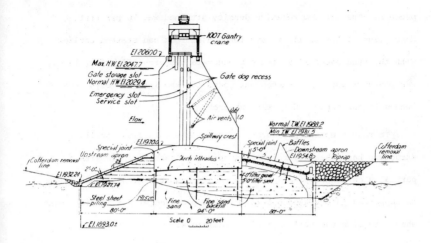

Fig. 10-Main Spillway-Transverse Section

The arch reactions are carried by the bedrock of the canyon walls. Upstream and downstream aprons with partial steel sheet piling cutoffs control uplift pressures, seepage flows and guard against foundation erosion.

Construction of the spillway structure required that the area be unwatered; the boulder layer overlying the alluvial sands and gravels be removed; the area be backfilled with sand compacted in the dry; the spillway structure be completed to above flood levels and the cofferdams removed; all in a single low-flow season. Delays and uncontrolled seepage resulted in the Contractor being permitted to place portions of the sand backfill under water. All sand backfill placed under water was compacted by means of the Vibrofloatation process. The minimum relative density attained was 70 per cent. Concreting of the spillway bays founded on rock had started earlier. With the sand backfill at grade a concrete sub-arch, five feet thick, was poured first. All subsequent pours for the interior bays were bonded to and supported by the sub-arch.

To reduce uplift pressures to acceptable values the spillway structure has a heavily reinforced upstream concrete apron with a partial steel sheet piling cutoff along its upstream end. This apron is supported on the sand backfill and is provided with a special joint where it joins the spillway weir. Additional rows of steel sheet piling were driven under the weir to eliminate direct seepage paths should the sand foundation settle with time. The rows of steel sheet piling do not provide a positive cutoff but their depth is sufficient to contain the foundation materials and impede the seepage flows.

To control the erosion downstream of the spillway structure a downstream sloping apron with baffles and a deflector lip was provided. The foundation soils at the toe of this apron are protected against erosion by another row of steel sheet piling. A filter and system of drains, similar to that described for Project B, under the downstream apron safely conducts seepage flows to a transverse drainage tunnel at the end of the apron.

Consolidation and curtain grouting was carried out in the rock of both abutments. The sand foundation, where the steel sheet piling terminated along the walls of the buried channel was also grouted upstream and downstream of each row of piling.

Piezometers for measuring the uplift pressures along the direction of flow were installed as were settlement devices for measuring the settlement of the sand backfill under the arch. Grout pipes were installed in the piers to permit grouting of any spaces if settlements did occur. To date no measurable settlements have been measured. A typical set of uplift pressure readings along the spillway is shown in Figure 11.

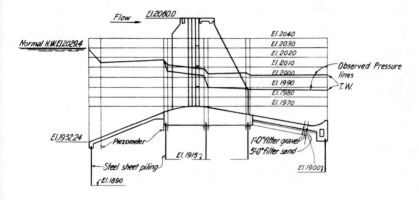

Fig. 11-Pressure Levels in Spillway Section

Inspections of the project in 1967 and 1972 revealed that there were no unfavorable changes in the rock forming the foundations and abutments. No settlement of the structures or separation of the sand backfill from the underside of the arch has been recorded. The upstream apron has been partially buried by sands and gravels carried in by the river. Scour below the dam has been that anticipated and has stabilized for the flows experienced to date. At 25 to 50 feet downstream of the sill scour had reached a depth of 10 to 12 feet below the crest of the

sill. At the sill no more than 10 feet of scour has been recorded.
The overall height of the sill is 19 feet.

CONCLUSION

Three early hydro-electric projects are described where the
foundation conditions at each site presented special problems.
The design and construction solutions to these problems are presented.
The successful performance and operation of the projects indicates
that hydraulic structures of significant size can be built and operated
where the foundation conditions are less than ideal.

References:

1. Project A - Harza Engineering Company Files.

2. Project B -

    - Fucik, E. M., "Petenwell Hydroelectric Project,"

      A.S.C.E. Transactions, Vol. 117, 1952, p. 528.

    - F.P.C. Inspection Reports, Harza Engineering Co., 1967, 1972.

3. Project C -

    - Guess, A. P., "Box Canyon Hydroelectric Project,"

      A.S.C.E., Journal of the Power Division, Vol. 84,

      No. PO3, Proc. Paper 1672, June, 1958.

    - F.P.C. Inspection Reports, Harza Engineering Co., 1967, 1972.

PROBLEMS ENCOUNTERED IN CONSTRUCTION OF DAM FOUNDATIONS

by

R. W. Bock,[1] W. G. Harber,[2] M. Arai[3]

Bureau of Reclamation dams are constructed on a variety of
foundations, nearly all of which are nonhomogeneous and anisotropic.
To properly determine the interrelationship of a dam and its foundation, Bureau engineers and geologists formulate geologic exploration
programs to define pertinent geologic features and discontinuities.

However, it's sometimes difficult to accurately define all
geologic discontinuities in a complex foundation and, during construction, engineers are frequently faced with unforeseen problems.

This paper discusses some of the problems encountered during
the construction of two earthfill and two concrete dams and the
solutions to those problems.

SOLDIER CREEK DAM

Soldier Creek Dam illustrates problems involved in bonding an
earthfill dam to an open-jointed, fractured rock foundation. Soldier
Creek Dam is located on the Strawberry River on the eastern slope of
the Wasatch Mountains about 75 miles southeast of Salt Lake City.
The dam is approximately 1,290 feet long at the crest and has a height
of approximately 251 feet above the bed of the river. It is a zoned

---

[1] Head, Earth Dams Section, [2] Supervisory Civil Engineer, and [3] Civil
Engineer, respectively, Bureau of Reclamation, Engineering and Research
Center, Denver, Colorado.

Soldier Creek Dam - Looking upstream

Soldier Creek Dam - Vertical joints in abutment
     foundation.

embankment with silty clay core supported by outer shells of sand
and gravel. Construction began in 1970 and was completed in 1973.

The damsite consists of a V-shaped valley which has been cut
through an up-thrown block of the Green River Formation. The Green
River Formation is typically thin-bedded, fine-textured lacustrine
deposits. It consists predominantly of fairly even and continuous
beds of siltstone and shale with some limestone and calcareous beds
of sandstone. Channels have been cut into these typical Green River
beds and have been backfilled with clean sandstone. The rock is
nearly all fairly hard; however, it is cut by horizontal bedding
at 1/2- to 24-inch intervals and two sets of vertical joints spaced
from 6 to 30 inches.

Preconstruction investigations indicated that the bedding and
joints in the Green River Formation were tight below river level;
but that open joints, tightening with depth, could be expected in
the upper 100 feet of the formation on the abutments. The formation was partially concealed by 0 to 30 feet of overburden.

From the preconstruction investigations, designers concluded
that it would be difficult and expensive to provide a water barrier
in the top of the abutment formation by grouting; consequently, the
specifications for the embankment included a 15-foot-deep, 30-foot-
wide key trench with side slopes of 3/4:1 excavated in rock in the
bottom of the cutoff trench on each abutment. Other foundation
treatments included a typical main grout curtain injected into the

Soldier Creek Dam - Grout cap trench.

formation through a grout cap in the bottom of the key trench and a pattern of secondary grout holes angled in the opposite direction from holes in the main curtain. The secondary grout holes were to be drilled and grouted after the completion of other grouting to check the effectiveness of the main grout curtain.

When the contractor excavated the 30-foot-wide key trench, cracks at the joints in the upper 15 feet of the formation were found to be wider than had been anticipated. In fact, the contractor was able to excavate the formation with a one-tooth ripper and a small amount of blasting. This method of excavation had loosened the fractured rock along the cut slopes, but it was apparent that many joints were open upstream and downstream from the key trenches which would provide an open seepage path directly to the trenches. Under these conditions a 30-foot-wide cutoff was not considered adequate to control the high heads in the lower portions of the abutments and an order for changes was issued to flare the bottom width of the key trench from 30 feet near crest level to 100 feet at river level.

Excavation for the grout cap was difficult due to overbreak in the loose, blocky rock. After several trial-and-error methods had failed to provide a satisfactory trench for the grout cap, an acceptable trench was obtained by presplitting the line holes using small charges of 40 percent powder taped, at 1-foot intervals, to prima-cord with the top charge located 18 inches below the collar of the hole.

Prior to placing grout cap concrete, grout pipe was placed for special holes located to intersect prominent joints that had been exposed in the grout cap trench. These special holes were in addition to those in the main curtain and were drilled and grouted prior to grouting the main curtain.

Grouting special holes and the main curtain holes was accomplished without difficulty. Grout takes were low, averaging about one-half cubic foot of grout per foot of hole.

The secondary grout holes were drilled normal to grout holes in the main curtain after the main curtain was complete. These secondary holes had very little water loss in water tests and accepted practically no grout indicating that a tight curtain had been obtained.

After the abutments were thoroughly cleaned and made ready for the embankment core, it was found that there were open joints in the formation extending from the surface into the abutments. These joints varied in width from about 1 to 6 inches and were up to 10 feet long. In addition, where the bottom of sandstone layers came in contact with shale, there were several places at which semicircular indentations of 8 to 12 inches occurred. One hundred and forty-nine of these joints and indentations were grouted. Grouting was accomplished by laying a 2-inch grout pipe in the joints and indentations. This pipe was extended up to the abutment by a riser pipe and sealed in the joint or indentation with specially compacted earthfill. A fill 20 feet deep was placed over the joints and indentations which were then grouted through the riser pipes.

The core earth fill next to the rock surface was compacted with hand-type compactors until a firm compaction could be made with roller embankment compactors.

TRINITY DAM

Trinity Dam illustrates stability problems that develop when foundation excavations are required in an abutment slope that is already near its critical angle of repose.

Trinity Dam is an earthfill dam located on the Trinity River in the Klamath Mountains of northern California about 200 miles north of San Francisco. The dam is approximately 2,450 feet long at the crest and has a height of 465 feet above the bed of the river. Its structural height of 537 feet above bedrock made it the world's tallest earthfill dam at the time of its construction in 1957-62. The dam consists of sand-silt-clay core supported by outer shells of sand and gravel and rockfill.

Bedrock underlying the Trinity damsite is comprised of two metamorphic formations. The riverbed and the major fraction of the canyon walls are cut in the Copley Meta-andesite Formation. This rock is the greenstone widely exposed in the Klamath Mountains and in the Trinity River Canyon walls. It is a hard crystalline rock resulting from low-grade metamorphism of basaltic and andesitic lavas,

Trinity Dam – Embankment buttresses against
left abutment slide area.

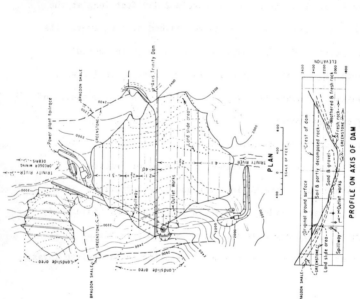

Trinity Dam – Landslide areas.

tuffs, breccias, and feeder dikes. The rock itself in an unweathered, unaltered form would be highly competent as a foundation but at the damsite it is extensively fractured, sheared, faulted, and weathered so that its basic characteristics no longer have a bearing on its competence as a foundation. Soundness is quite variable over the foundation depending on the degree of alteration that has taken place.

Minor fractions of the abutments and higher canyon walls are cut through the Bragdon Formation. That formation consists of siliceous and graphitic shales, sandstones, and conglomerates. The shales are brittle, thin bedded and hard when siliceous; and weak, broken, and very fissile when graphitic. The conglomerates and sandstones, roughly 25 percent of the formation, are hard, tough to brittle, and generally thin bedded.

Materials overlying bedrock include (1) dredged and undredged streambed gravels averaging 30 feet in depth, covering narrow water-worn channels or plunge basins that cut into rock to twice that depth, (2) areas of landslide detritus and slope wash resting at stream level and partially overlying the streambed gravels, (3) intermittent areas of landslide detritus and weathered rock blanketing oversteepened canyon walls to a depth of several hundred feet and representing active or potential slide areas, and (4) weathered rock in place.

Landslides on lower canyon slopes are a common feature of the Trinity area because the walls along most streams are oversteepened. The landslides are generally shallow and sliding apparently takes place

on several spoon-shaped surfaces in the upper part of the slide and as
a mudflow in the lower part. Three such landslides were identified
in or adjacent to the damsite by preconstruction investigations.

One slump or landslide deposit in the foundation area was located
on the right abutment upstream, partially displacing and overriding
streambed gravels. The base of these debris, at and below stream
level, was a soft, silty, gravelly, clayey mass. The top-back slope
of the slide was backfilled with a loose rubble. This slump deposit
was removed by about 1,000,000 cubic yards of foundation excavation.

A second similar but more extensive landslide area was identified
on the left abutment, downstream. Though beyond the limits of the
dam, further movement of the mass would have endangered the spillway
stilling basin. Since the material composing this slide mass was
found suitable and economically desirable for use in the embankment,
it was decided to unload the slide rather than relocate the spillway.
Nearly 5,000,000 cubic yards were borrowed from this area.

The toe of a third potential slump mass is present under the upper
half of the left abutment foundation in the area of the outlet and
spillway shafts. This bank of debris showed no sign of recent movement and would appear to be buttressed by the dam embankment. Since
maximum distress would occur during construction, it was decided to
apply remedial measures here if and when they proved necessary. As
a result of this decision, a number of problems developed during the
construction period.

The fault band under the left abutment slump mass was exposed early in the construction period by excavations for the dam foundation and excavations for the spillway and outlet work, shafts. It was about 1 foot wide and consisted of black clayey material.

In the fall of 1958 when the embankment was more than 100 feet below the toe of the slump mass some slippage took place. Projections of the upper surface of the shear zone of from 2 to 4 inches out over the adjacent lower part in the face of excavations constituted confirmation of the slippage and represented roughly the maximum amount of actual movement. Apparently a quarter to half a million cubic yards had just started to move. The contractor was ordered to remove some temporary stockpiles of excavations he had placed on the slide area and to discontinue excavation and grouting operations near the toe of the slide. No further movement was detected in the 1958 construction season.

In preparation for the 1958-59 winter rain season surface cracks in the slide area were sealed and drainage trenches were cut from all depressions in the left abutment so that water could not collect on the abutment. Despite these precautions, moisture penetrated to the slide plane resulting in another inch or two of movement in January and about a foot in February. With the arrival of the 1959 spring-summer dry season, movement stopped. At this time it was decided to postpone foundation treatment through the slide mass until the embankment was constructed high enough to support the slide. Treatment

would then consist of thoroughly grouting the shear zone and the disturbed rock and slump material above the shear zone in a 100-foot-wide
band adjacent to the axis of the dam.

By fall of 1959 the fill had been brought to a level about 20 feet
below the toe of the slide and it was evident that treatment of the
slide area probably could not be completed before the rainy season if
normal placement operations were continued. Rather than risk reacti-
vation of the slide by wet weather the contractor was directed to
buttress the abutment by placing mounds in the outer zones of the
upstream and downstream sides of the left side of the embankment.
These mounds were constructed ahead of the remainder of the embankment
and were approximately 70 feet high, 200 feet long and 300 feet wide.
When the buttresses were in place, the foundation area, 100 feet wide,
between the buttresses, was grouted through overburden down to sound
rock below the shear zone. After grouting was complete, overburden
was removed as far as necessary to reach a horizon where voids were
filled with grout. The special abutment treatment was completed in
November 1959 without damage to the spillway and outlet works shafts.
No further movement of the left abutment slide was noted.

Trinity Dam was completed in 1962 and has been filled several
times. No difficulties have been experienced in the left abutment
slide area from further movement or from seepage.

PUEBLO DAM

Pueblo Dam is being constructed on the Arkansas River about
6 miles west of Pueblo, Colorado. When completed, the dam will be
a concrete and earthfill structure about 10,200 feet long. The
central concrete portion will be about 1,750 feet long and will consist of an ungated, 550-foot-wide overflow spillway section and non-
overflow sections flanking the spillway. In general, each nonoverflow
buttress head is 75 feet in width, with the buttress 18 feet wide.
The upstream slope is 0.4 to 1 and the downstream slope is 0.45 to 1.
These dimensions vary slightly in the spillway section of the dam.

This central concrete portion will be the first massive head
buttress dam designed and constructed by the Bureau of Reclamation.
Its 180-foot height also places it among the highest of its type in
the world.

The construction of the dam was divided into two separate contracts. Implementation of the first contract was begun in July 1970
and completed in December 1971. Work under the second contract was
begun in July 1972 and is scheduled to be completed in September 1975.

The first contract was essentially an excavation contract while
the second was a structures construction contract. The structures
construction contract was awarded only after the foundation was excavated to final grade. With this approach, the Bureau of Reclamation

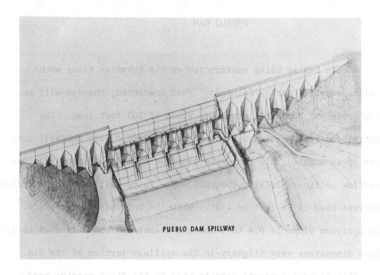

PUEBLO DAM SPILLWAY

Pueblo Dam – Artist's sketch of Pueblo Dam, concrete
   dam portion.

Pueblo Dam – Aerial view looking east at the excavation
   for the buttresses, the spillway stilling basin, and
   outlet channel.

Pueblo Dam - Aerial view looking south along the
concrete dam axis at the foundation excavation
for the buttresses and the spillway stilling
basin.

hoped to eliminate structure construction delays which frequently
occur when unexpected geologic discontinuities within a foundation
are uncovered.

The damsite is located in a wide, flat valley; the present river
channel is located in a gorge about 75 feet deep by 100 feet wide

The Dakota and Purgatorie geologic Formations underlie the concrete portion of Pueblo Dam. The surface rock is Dakota sandstone
which is approximately 90 feet thick. Included in this sandstone are
a number of discontinuous layers and lenses (horizontal to near horizontal) of shale and carbonaceous laminae. Underlying the Dakota
sandstone is about 75 feet of Glencairn shale and about 100 feet of
Lytle sandstone both of which comprise the Purgatorie Formation.

Excavation for the buttress dam was laid out to leave as much
sandstone above the Glencairn shale as practical. An inclined
(upstream dipping) rock foundation base for individual buttresses
was adopted, since it will improve the sliding and shear stability
in the buttress foundation. The slope of the base was approximately
4 percent. Also, the excavation was laid out to follow the shape of
the buttress so the stresses from the dam and reservoir will be better
distributed to the rock foundation. Except in the river gorge area,
approximately 70 feet of Dakota sandstone remains under the dam. In
the river gorge area, portions of the base of the concrete are in contact with the Glencairn shale.

A concrete gravity plug was constructed in the river gorge area. Its crest was level with the excavation for the adjacent buttresses. During excavation for the concrete plug, the contractor encountered an unexpected site condition at the right abutment near the upstream end of the concrete plug. The preconstruction geology report had indicated that on the right side of the river, there were three sets of joints on the surface, all of which were tight on the surface.

As excavation proceeded with the abutment, two of the joint systems became more open and formed an area of highly fractured and broken rock. The contractor was directed to reslope and remove the fragmented rock, resulting in an additional 1,000 cubic yards of excavation. To further enhance the stability in this area, the depth of consolidation grouting was extended from 30 feet to between 45 and 50 feet.

When the contractor reached the bottom of the plug excavation, the material expected in the bottom was not sound sandstone but rather consisted of a highly weathered shale subject to rapid deterioration when exposed to air and water. An additional 1,000 cubic yards of material outside the original excavation line was removed from this area and the bottom elevation was lowered by almost 5 feet.

The original construction drawing specified that consolidation grouting would be performed prior to placing concrete. However, the shale foundation material was unable to provide a seal for the grout

under even the lowest grouting pressures. It was decided that a minimum of 10 feet of concrete would be placed on the foundation before
permitting grouting. The concrete "cap" performed as anticipated and
satisfactory foundation consolidation grouting was obtained.

By October 1971, excavation for the buttresses was virtually
completed to the final grade shown on the construction drawings.
However, as the excavation was being completed, it became obvious
that shale seams and carbonaceous laminae of considerable continuity
existed several feet below the anticipated final excavation grade.
These were known to have existed at random locations in the foundation
from earlier geologic exploratory drilling but their degree of continuity and precise location could not be determined.

The most critical area was from Buttress 8 through Buttress 14,
the spillway section, where downstream rock support was removed to
allow for construction of the spillway stilling basin. For example,
from visible observation a relatively large shale seam underlined a
portion of the Buttress 14 foundation. If any extensive and continuous carbonaceous laminae or shale seams underlied the excavated rock
surface, it would result in lowering the sliding factor of safety
of the individual buttresses. In order to preclude such a condition and to better understand the requirement for any additional
excavation, additional exploratory drilling was performed immediately before and just after award of the structures construction
contract. The exploratory holes at 30-foot centers along the

centerline of each of the spillway buttresses were drilled to determine the competence of the foundation. As a result of the investigation, a total of approximately 1,300 cubic yards of additional
rock was removed from Buttresses 8, 10, 11, 12, and 14 under the
structures construction contract.

The contractor's operations under the structures construction
contract were considerably curtailed during the late fall and winter of 1972 due to unusually severe weather conditions. Accordingly,
the contractor was unable to start his concrete placement until
April 1973. During this time there was some apparent surface deterioration in the excavated rock surfaces, as manifested by leaching
of carbonaceous laminae, air and water slaking of exposed shale
seams, and drummy areas of sandstone. The latter probably occurred
as a result of stress relief movement during excavation. Freeze-
thaw action also could have been responsible for some of the drummy
layers. Prior to concrete placement, the entire foundation was
examined for drummy rock. Since it was impractical to remove each
small localized area of disbonded rock, the contractor was instructed
to remove only those zones having an areal extent of 20 square feet
or greater. Approximately 20,000 square feet of such rock area were
removed from various locations in foundation of the buttresses. Maximum depth of drummy areas was 12 inches plus or minus. After completion of this operation, the foundation was determined suitable for
placement of concrete.

THIRD POWERPLANT AT GRAND COULEE DAM

Grand Coulee Third Powerplant, which will have an initial generation capacity of 3,900 megawatts, is located approximately parallel to the right bank of the Columbia River immediately below Grand Coulee Dam in an area previously occupied by the old right abutment switchyard. Structures to be constructed include a forebay dam and forebay, powerplant, powerplant switchyard, and visitor's center. Modification of the two existing powerplants and the left abutment switchyard was required to accommodate the new structures.

The forebay dam, a concrete gravity-type structure, is located along the right abutment at approximately a right-angle extension of the present dam. The dam is about 1,200 feet long with a height of about 200 feet above the forebay floor, and has an upstream vertical face and a downstream face with a slope of 0.65:1, horizontal to vertical. The crest of the dam is 30 feet thick. Six 40-foot-diameter penstocks will supply water from the forebay dam to the powerplant. The dam is constructed on a bench approximately 160 feet above the river level and 100 feet above the main floor of the powerplant. This bench serves as the dam foundation as well as the floor of the forebay channel.

As was done later on Pueblo Dam, as previously discussed, the Bureau divided the contract work basically into two parts: excavation and structures construction. This division was made to avoid delays

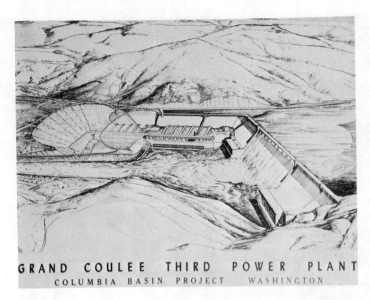

Grand Coulee Third Powerplant - Artist's sketch of Grand Coulee Third Powerplant.

Grand Coulee Third Powerplant - General view showing construction progress of the forebay dam and powerplant.

Grand Coulee Third Powerplant - Aerial view showing the
east end of existing Grand Coulee Dam and forebay
channel excavation. Note some portion of the
shoulder for the road was lost due to the jointing in
the rock.

Grand Coulee Third Powerplant - Aerial view showing the forebay channel excavation.

Grand Coulee Third Powerplant - Aerial view of the upstream cofferdam showing excavation activities for removal of the cofferdam. Note retaining wall at the saddle area near the central reach of the cofferdam.

in structures construction which could be caused by special treatment and extra excavation required by unexpected faults and shear zones in the foundation rock.

The work was divided into two excavation contracts and a prime contract for completion of the excavation and the construction of the dam and powerplant. The first excavation contract which included the excavation of the forebay channel, a portion of the dam foundation, and construction of the cofferdam was begun December 1967, and completed in September 1969. The second excavation contract, which included the excavation for the remainder of the dam foundation and a portion of the powerplant, was started in August 1968, and completed in July 1970. The prime contract was begun in March 1970, and is scheduled for completion in January 1976.

The granitic rocks which comprise the right bank abutment of existing Grand Coulee Dam are durable and of good quality. Two types of granites are found. The older granite is medium to coarse grained, massive and rough surfaced, and has weathered by exfoliation to large rounded bluffs and knobs. Most of the jointing in the older granite is widely spaced and has not greatly affected the strength of the rock. The younger granite has intruded the older granite generally north-south, with near vertical dike-like bodies. The younger granite is fine grained and is cut by a myriad of closely spaced randomly oriented joints.

During excavation for the Marina Way Roadway, the left shoulder
of the road, adjacent to the forebay channel, was lost due to the
jointing in rock. To maintain the proper width of roadway, a concrete retaining wall along the left shoulder was constructed as part
of the work under the prime contract.

The condition of existing joints in the foundation rock is very
critical to the stability of the forebay dam. Bureau engineers were
concerned that even normal blasting could open existing joints and
make them more continuous. Accordingly, additional blasting restrictions were proposed to maintain the integrity of the joints. The
restrictions were accomplished by reducing the bit size of blasting
holes and the powder factor, by close spacing the drill hole pattern,
by decking blast holes and delaying the decks separately, and by
limiting the maximum amount of powder per delay.

To accomplish the excavation for the forebay and forebay dam,
a cofferdam consisting of a "left in place" rock barrier and a steel
sheet piling structure was constructed upstream from the existing
Grand Coulee Dam. The rock portion of the cofferdam had a crest
with an average width of 40 feet, a height of 190 feet above the
forebay floor and upstream and downstream slopes of 3/4:1 horizontal
to vertical.

The initial geological investigations indicated that the rock
cofferdam would be located in sound rock and that the rock surface
was fairly uniform. However, during excavation of the upstream and

downstream faces of the cofferdam, a deep saddle area, approximately 40 feet deep by 60 feet long, was found near the central reach of the cofferdam. To bring the rock cofferdam to grade, retaining walls were constructed at the upstream and downstream sides of the saddle area and the space between the retaining walls was filled with earth materials.

The geologic investigations of the forebay dam foundation revealed two major faults. These were extensively studied to determine their location and size and to evaluate the extent of treatment. One fault occurred under the forebay dam and ran nearly parallel with the axis of the dam. As initially defined, the fault dipped approximately 60° downstream and had an apparent width of 10 feet plus or minus. Bureau engineers decided that treatment of the fault was necessary to maintain proper foundation stability. This treatment consisted of excavating and placing backfill concrete in the fault zone to a depth of 25 feet. However, as excavation progressed, it became apparent that about 150 feet of the fault near the central reach of the forebay dam was considerably wider (50 feet) than originally estimated. In this area, there was a distinct independent gouge zone on each side of the fault separated by ribs of hard but very jointed granite. The entire 50-foot width of the fault was excavated to a depth of 25 feet and backfilled with concrete. The other fault approximately normal to the axis of dam did not require any special treatment.

As part of the treatment for the fault, a series of near horizontal drains were drilled upstream from the sloping rock surface
below the toe of the dam. These holes originated near the base of
the slope and were oriented to intercept the fault. The horizontal
drains were designed to reduce pressure against the gouge material
in the fault and also reduce uplift pressures in the foundation for
the dam. As more information on the characteristics of the fault
zone became available, it was recognized that the gouge material
was very susceptible to piping. To prevent piping and plugging of
the drain holes, approximately 1,100 feet of slotted polyvinyl-
chloride pipe was installed in the drains where they crossed the
gougy area of the fault. The slots were sized to prevent gouge
material from piping yet not impede water migration from the fault.

PROBLEMS RELATED TO THE CONSTRUCTION OF

DAM FOUNDATIONS FROM A CONTRACTORS' VIEWPOINT

by John W. Leonard*

There are two types of dam foundations from the Contractors' viewpoint. These two types are good foundations and bad foundations. Good foundations are those a Contractor makes money on and bad foundations are those he loses money on.

WHAT MAKES GOOD FOUNDATIONS? (IN SIX PARTS)

1. Proper and adequate information for the Contractor to judge the character of work, plus specs and drawings endowed with clarity to define the scope of work.

    What is adequate information for the Contractor to judge the character of the work? This would include standard penetration tests, unconfined compression tests, shear vane tests, regional and local geological reports, permeability and pump tests and other similar information. Permeability and pump tests will help decide on the type of unwatering pumps to be used, will help to arrive at the number of well points, etc. The blow counts will indicate the stability of the excavated slopes. The logs of drill holes along with the physical layout will describe chances of piping and any slope stability problems. The shear vane tests, blow counts or unconfined compression tests will help the Contractor decide what type of equipment to be used. It might indicate that only twin scrapers can be used and/or draglines on mats. Geological logs of the drill holes might indicate a series of stratas that would not allow successful use of deep wells and/or well points. The geological report, particularly the regional, will help the Contractor to make his decisions as to what the various information, including his site visit, is telling him. His solution should be consistent with the overall geological background.

    As I look around here today, I see in the audience several members of the industry we have used from time to time in pre-bid stages to check our conclusions, and they, of course, have used this information. We have not always followed their recommendations, but generally we have sat

*Chief Engineer, Morrison-Knudsen, Boise, Idaho

115

down, thrashed out our differences of opinions and reached a mutual, satisfactory conclusion.

There is much information developed for design that can be used by the Contractor and/or his consultant and should be made available to him in the pre-bid stage.

2.   Alternate provisions in the contract and specifications, and administration thereof to cope with unforeseen conditions in a reasonable manner. This spans beyond contractural matters into engineering. (Covered in 3 below)

It is difficult or almost impossible, and probably the best word is impossible, to write specifications to include all the many things that might happen with a foundation. Much of the foundation work being accomplished has been performed for governmental bodies or agencies. This is why Federal and State contracts have the Article IV Provision incorporated into them. The contract-legal world has fairly well settled down with many decisions, rules, etc. revolving about Article IV. Quoting from it:

(1) Subsurface or latent physical conditions at the site which differing materially from those indicated in the contract, or (2) Unknown physical conditions at the site, of an unusual nature, differing materially from those ordinarily encountered and generally recognized as inhering in work of the character provided for in this contract.

This recognized the possibility of unusual conditions. We can expect them and the machinery is set up to solve them.

3.   Practical engineering solutions of new design problems often uncovered in the process of excavation. Timeliness is important here.

A previous session was held at Asilomar, May 14-18, 1972, Section 6, subject "Owner, Engineer, Contractor Economic Considerations in Dam Construction", Chairman being the honorable J. P. Buehler. The opening speaker was George H. Atkinson, Chairman of the Guy F. Atkinson. Mr. Atkinson stressed several times that "time is money, not just overhead, but ongoing daily expenses such as job overhead, shop crews, maintenance crews, light and power and standby expenses of various kinds — a considerable total cost for every day the job is prolonged." I am sure Mr. Atkinson had partly in mind time lost with foundation problems. At the critical point in the foundation work, serious consideration should be given to providing on the site a short term capability of design personnel who working with the engineers' experienced field supervision personnel can quickly resolve the foundation problems that arise. There is no substitute for seeing the problem and discussing it at the site. This leads to early resolution. Granted there may be calculations in design to be performed, but the difference between one answer and another is generally a significant difference and does not involve long time detailed calculations.

4. Part of a foundation is grouting. We have several comments on grouting, and I think most of these are pretty well understood; they have been pretty well resolved over the years, but probably won't hurt to bring them up again.

It is well for the engineer forces on inspection of the site to develop a good relationship with the Contractors' grouting people. They can be helpful. They are knowledgeable themselves. Grouting is a gamble for a Contractor, particularly on quantity. There should be no resistance in letting them make a dollar. Grouting is not all that expensive. Don't clutch up when a hole starts to take.

Grout galleries are a good thing. It permits one to go ahead with the main job without being held up. Also, they provide something to grout against. You would be surprised how many times it is difficult to get inspection forces to realize the difficulties in surface grouting consolidation. Most Contractors have experienced being required to literally grout a haystack.

5. One of the problems we have is unwatering. How dry is dry? Part of the concern in unwatering is the possible loss of fines in the foundations. This often is usually not a major problem. It looks a lot worse than it is. Water seeping in the foundation appears muddy, but less than one-half percent of fines will make it appear quite muddy. Any experienced well driller knows the loss of fines extends back a very short distance based on how they use their well screens. So a loss of fines will extend back only a few feet in the foundations. Some specs

require deep wells in the foundation. Many foundation of non-dam structures are never unwatered, they are tremied, ten percent extra cement is added. So, in a dam foundation if there is some leakage, maybe extra cement is indicated. Earthfill can be placed over optimum moisture on rock foundations. It is more plastic and can be well compacted.

6.   Designs that consider todays' economic conditions.

No Contractor has yet to make money as he reaches final grade in rock, does the clean-up and starts placing the bottom of the structure on the foundation, be the structure dirt or concrete.

This is one reason why slurry trenches have proved economical. One does not have to get down and clean the foundation by hand. I am thinking here of the deep-type cut-off trench 70 to 90 feet deep that can be dug with equipment, cleaned up with equipment and then backfilled with material taken from the ditch or with tremie concrete. This slurry also works well with intermediate types of foundations such as a weak sandstone that deteriorates under running water and construction equipment. This is a design that considers today's economical conditions.

Other considerations: Many specifications provide payment for secondary foundation clean-up, for final foundation clean-up and excavation. This is a good thing. It recognizes that the engineer may want to inspect the foundation and continue on. The primary clean-up for inspection is not to be the final clean-up. It puts the engineer and the Contractor in a better relationship. Along the same line are provisions for dental concrete and dental excavation.

I would like to congratulate the engineering profession on their handling of safety features. I think the more and more prevalent idea of paying by unit prices for safety provisions such as tunnel supports, shotcrete, anchors, mesh, gunite, etc. is excellent. It is just not a good thing to have the Contractor absorb in a lump sum any safety provisions that may be required. It is an invitation to gamble with the safety of people and, in some cases, the project cost. Each Contractor is on the same bidding basis.

The increase in construction wages in the last five years has been nothing short of phenomenal. Some construction practices that have built up over the years were started and developed in periods of low wage rates. Here, I am thinking of the clean-up specifications. Rock foundations are often cleaned up where one could literally eat off the rock surface; all the crevices are cleaned out — sometimes with 16 penny nails and 10 quart pails. At a dollar an hour that is one thing and it doesn't make all the difference; at 20 dollars an hour is there that much benefit? As I remember from school, we used to play with the angle that an inclined surface makes with a horizontal surface. The tangent of this angle has a direct relation with the friction angle. One placed a book on an inclined plane (another book) and then kept increasing the angle until the top book slid. That angle is the friction angle. At school the good text book said for sliding of concrete on rock MU was about .5. Looking at these two books that angle would be about .5; we are doing a hell of a lot of MU at 10 to 20 dollars an hour.

We have costs on concrete dams' foundation clean-up; they are unbelieveable. We have costs of two man hours per yard (one plus man hour is about average). This is highest case, but it covered a lot of yards. You can carpet a house with very fine carpet for a man hour per square yard today.

WHAT MAKES BAD FOUNDATIONS? (IN THREE PARTS)

1. Proper and adequate information is not supplied to the Contractor. The bad news is hidden.

2. Phases in the specs such as:

    "Pre-construction conference prior to excavation of the core trench and constructing the main dam embankment: The Contractor and his employees or sub-contractors to be assigned responsibility for carrying out the work shall meet at the project office with representatives of the Contracting Officer who designed the earth embankments to discuss requirements of the work to be done. All aspects of the construction will be discussed including unwatering requirements, stream diversion and closure, placement requirements and the Contractor's proposed plan of operation. This conference is required in order to develop mutual understanding between the Contractor and the Contracting Officer about important or key phases of project construction."

How does the Contractor price this in his bid? How do we allow in our bid for the requirements that may be brought up in such a meeting? Items of placement requirements and other things of that nature that are not already in the specs. What are they? The specifications permit the Contractor and the engineer to develop study and to arrive at price for the work outlined. We have no opportunity to price what comes out of such a post-bid conference.

3.   Provisions that put unforeseen conditions on the Contractor.

"Embankment stability during construction can be affected by pore pressures induced by the weight of fill placed. The development of pore water pressures during construction in either the foundation or in the embankment will be observed by Government forces reading various installed instruments as described elsewhere in these specifications. Evaluations of embankment and foundation stability will be made during construction by the Government, and the Contracting Officer reserves the right to decrease the rate of fill placement or temporarily discontinue fill operations as necessary for the stability of the structure." This is not fair. Why should the Contractor be asked to absorb in his bid the risk of a slow rate of placement or even discontinuing the operations.

SUMMARY

I appreciate the invitation to present this paper. The remarks herein are intended to be helpful and if they appear to be critical, please extend the blanket of constructive criticism over them. Dam foundations can be difficult for both of us. Here's to more good foundations.

FOUNDATION CONSTRUCTION PROBLEMS AT
CORPS OF ENGINEERS PROJECTS

Paul R. Fisher,[1] John C. Bowman, Jr.,[2] Benjamin I. Kelly, Jr.[3]

INTRODUCTION

For the purpose of this paper, a foundation construction problem is
defined as a situation, occurring after completion of design, which
requires a change in foundation construction.

Case histories are presented in which the foundation problems encountered
required solutions that, for the Corps, were somewhat out of the ordinary.

The first case history describes a potential abutment seepage problem
which required the construction of a membrane wall within the abutment to
control seepage and assure adequate protection for the dam embankment
against piping and base erosion.

The second case history describes a foundation stability problem for the
spillway section of a major lock and dam where foundation grades could
not be lowered for fear of endangering the cofferdam. Here, cast in
place, reinforced, concrete caissons were used.

ABUTMENT SEEPAGE PROBLEM

Gathright Lake Project is under construction by the Corps of Engineers
Norfolk District. It is a multi-purpose project on the Jackson River
for flood control, water quality improvement, and recreation. The
project is located in Western Virginia about 180 miles southwest from

[1]Office, Chief of Engineers, [2]Norfolk District, Corps of Engineers,
[3]Louisville District, Corps of Engineers

Washington, D.C. Figure 1 shows the location of the dam outlet works,
and spillway with respect to the river and major geologic structures.
The dam will consist of a 257 foot high, 1,270 foot long rolled rockfill embankment with an impervious core.

A tunnel outlet works has been driven through the right abutment and a
remote emergency spillway will be located about 2-1/2 miles south of
the dam. The planned embankment zoning and foundation treatment is
shown on Figure 2.

The planned foundation treatment consists of:

     1. A multiline grout curtain with a minimum of three lines,
inclined in plane of the curtain (two directions) to a depth of
150 feet.

     2. Stripping to rock, shaping and dental work under entire
foundation.

     3. Consolidation grouting to 30 feet under impervious core and
filter zones.

The project is located in a terrain of valleys and ridges formed in
folded paleozoic sediments. The Jackson River flows southwest, down
a synclinal trough, in shales and then turns easterly and breaches two
anticlinal ridges of sandstones and limestones, forming the gorge in
which the dam site is located (see Figure 1). Below the gorge the
river turns southwest again and flows down another synclinal trough.
The western fold is a narrow anticline which forms a long narrow nose.
A normal fault is located on the anticline's east flank paralleling the

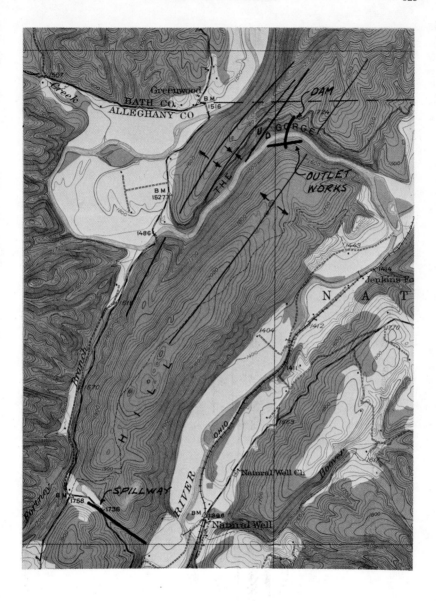

FIGURE 1. Location of Gathright Dam and Appurtenances with Major Geologic
Structures.

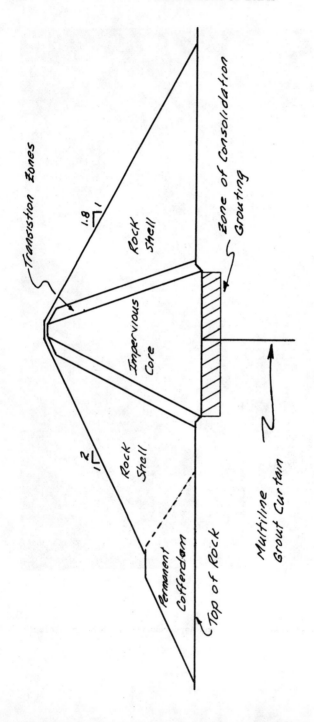

FIGURE 2.   Embankment Zones and Foundation Treatment

regional fold axes. The eastern fold is a broad anticline. Most of the
gorge is cut in this fold.

Figure 3 describes the lithology occurring in the project vicinity.
Figure 4 shows the sequence of formations in the dam's left abutment.

The Tonoloway, Lower Keyser, and Clifton Forge formations do not exhibit
significant solutioning in the vicinity of the dam site. The Upper
Keyser, Coeymans, and Licking Creek formations are solutioned to varying
degrees. Solutioning, occurring in the Coeymans formation, has produced
conditions conducive to leaching of the Healing Springs formation.

During design it was realized that two potential problems existed with
respect to the foundation. First, cavernous conditions were known to
exist in the limestones in the region - the possible occurrence of these
conditions in the reservoir and dam foundation needed to be checked out.
Second, the adverse dip of bedding in the upstream right abutment might
produce problems with respect to the integrity of the intake tower.

To avoid "taking chances" with filled or unfilled solution channels in
rock and to define the possible adverse geology problem on the right
abutment, a pre-construction abutment stripping contract was conducted in
1967 and 1968. The abutment stripping rather clearly defined the
geologic structural conditions on the right abutment. It did not so
clearly define the degree of severity of the problem on the left
abutment. A small cave had been uncovered but ubiquitous solutioning

NAME	THICKNESS	DESCRIPTION
LICKING CREEK	140'±	Soft to moderately hard, shaley limestone; variably weathered. Lower 80' is blue solutioned limestone.
HEALING SPRINGS	30'±	Hard, fine to medium grained, gray, quartzitic sandstone; moderate to highly jointed and fractured; lower portion is calcareous and leached.
COEYMANS	40'-60'	Hard, coarsely crystalline, gray limestone; closely jointed and fractured; solutioned & cavernous; upper part is sandy.
UPPER KEYSER	20'±	Hard, fine to medium grained, sandy, argillaceous limestone; moderately to highly fractured, somewhat solutioned.
CLIFTON FORGE	100'±	Hard, fine grained, gray, calcareous sandstone, grading to sandy limestone; medium bedded to massive; contains argillaceous zones.
LOWER KEYSER	60'±	Hard, fine to medium grained, gray, argillaceous limestone; medium bedded to massive.
TONOLOWAY	100'±	Moderately hard to hard, dark gray calcareous shale and argillaceous limestone; thin to medium bedded, contains fissile zones.

FIGURE 3. Geologic Formations Occurring at Gathright Dam.

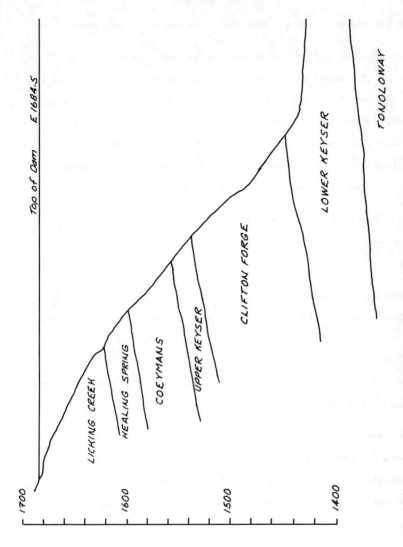

FIGURE 4. Location of Geologic Formations in Gathright Dam Left Abutment

was not present. Figure 5 shows the cave found on the left abutment.

During the period from 1969 to mid 1970, the left abutment was the
subject of a number of conferences. The consensus was developed that a
fairly deep core trench into the Coeymans formation together with a
multiline grout curtain would take care of any encountered conditions.
A minor re-alignment of the dam axis had been proposed and it was
decided to study this re-alignment and further define any solutioning in
the left abutment with additional explorations.

By May 1971, ten more borings had been drilled on the left abutment,
some seismic explorations had been done, and the Norfolk District was
convinced that they had significant solutioning in the abutment.

A special foundation investigation program to resolve this question
which included drilling 24 additional borings, using a TV camera in some
of these borings, and additional seismic studies. In addition, a board
of consultants was convened to assist in the analysis of conditions in
the left abutment and in determining the most applicable remedial measures.
Figure 6 is a borehole TV view of an encountered cave.

The board of consultants reviewed the information from the special
foundation investigation and proposed that adits be driven into both
abutments. These adits would serve three purposes. They would provide
the specific information needed to design the remedial treatment, would
serve as drain tunnels, and could serve as grout tunnels.

FIGURE 5. Cave Found on Gathright Dam Left Abutment After Stripping

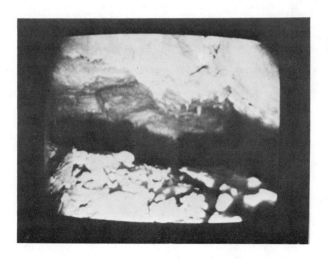

FIGURE 6. Borehole TV View of Small Cave in Gathright Dam Left Abutment

The recommended adits were driven in 1972. Three adits were driven in the left abutment, totalling 1,625 feet and two adits were driven in the right abutment, totalling 1,337 feet. In addition, intersecting drainage adits were driven in both abutments to provide permanent access and drainage.

Extensive solutioning was not encountered in the right abutment and it was concluded the grouting and drainage would be adequate treatment.

Extensive solutioning was encountered in the left abutment. Figures 7 and 8 are views inside one encountered cave. In addition to open caves, numerous solution widened joints were encountered which were partially filled with rock debris and clay. It was concluded that a concrete membrane cutoff along with drainage tunnels immediately downstream from the membrane were required to control seepage through the left abutment. The membrane will be eight feet wide, will have a maximum height of 107 feet, and will extend 730 feet into the abutment. The layout of the membrane is shown on Figure 9.

The bottom of the membrane will be embedded three feet into the Clifton Forge formation. It will run into the abutment parallel to the formation strike for about 400 feet and will then turn and follow the top of the Clifton Forge to the crest of the anticline downstream from the dam axis. It will form a cutoff up to elevation 1630, 6 feet above standard project flood.

FIGURE 7. View of Interior of Cave in Gathright Dam
Left Abutment

FIGURE 8. View In Another Direction of Same Cave
as in Figure 7.

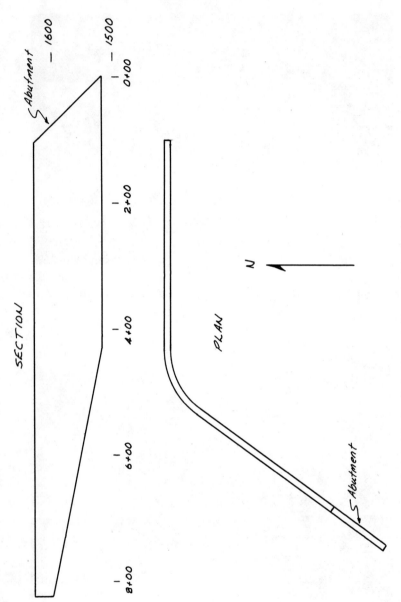

FIGURE 9. Layout of Gathright Dam Left Abutment Concrete Cutoff Wall.

The wall construction was included in the main dam contract and the wall
is currently under construction.

FOUNDATION STABILITY PROBLEM

Uniontown Locks and Dam is a major navigation project on the Ohio River
under construction by the Louisville District. It is located about one
mile upstream from the confluence of the Ohio and Wabash Rivers.

As shown on Figure 10, the project consists of two locks on the Indiana
bank, one 600 feet and one 1,200 feet long, a gated spillway section,
and a cellular fixed weir.

The rocks occurring at the site are sandstones, shales, limestones, and
thin coals of the Pennsylvanian Lisman formation. A description of these
rocks is shown on Figure 11. These sedimentary rocks are essentially
flat-lying but local flexures and fault zones exist.

Early design envisioned that the lock and spillway piers would be founded
on the shale below coal No. 1. This concept was revised during feature
design for the dam, partially based on foundation observations during lock
construction. The revised concept would found the dam piers in shales 2
and 3.

By early 1971 lock construction was complete, and the first stage main
dam cofferdam had been constructed and dewatered. Figure 12 is a view of
the cofferdam layout.

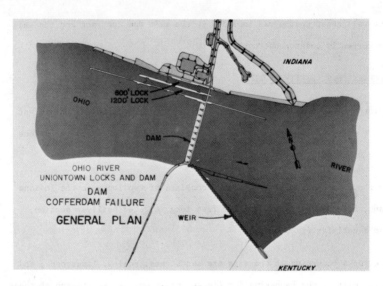

FIGURE 10. General Plan of Uniontown Locks and Dam

GRAPHIC COLUMN	ROCK UNIT	THICKNESS	DESCRIPTION
	MADISONVILLE LIMESTONE MEMBER	0 - 8.0' ±	LT. GRAY, HARD, ARGILL. TO SHALY, FOS. NODULAR W/WAVY BEDDING, THICK BEDDED TO MASSIVE.
	COAL, SHALE W U.C. (NO. 1)	0.8' - 2.0' ±	DK. GRAY TO BLACK, HIGHLY CARBO. W/COAL, NUMEROUS SLICKENSIDES, U.C. - GRAY SOFT SLICKENSIDES.
	SHALE	9' - 14' ±	MED. DARK GRAY, SILTY, SANDY, FAINTLY LAM. W/SOME INCIPIENT FRACTURING, PLATY TO SUB-FISSILE, MOD. HARD TO MOD. SOFT.
	SHALE	5' ±	DK. GRAY, MOD. SOFT TO MOD. HARD, SUB-FISSILE TO FISSLE, CARBO., SILTY, W/OCC. PHOSPHATE CONCRETIONS.
	SHALE W COAL & U.C. (NO. 2)	1.8' ±	BLACK, SOFT, VERY CARBO. TO IMPURE COAL, FISSILE, WAXY, CLAYEY IN PART.
	SHALE (NO. 1)	10' - 12' ±	GREENISH-GRAY, MOD. HARD TO MOD. SOFT, SILTY, SANDY W/NUMEROUS PARTINGS OF FINE SANDSTONE W/OCC. LS. NODULES, THIN BEDDED TO MASSIVE PYRITIFEROUS.
	SHALE (NO. 2)	15 - 20' ±	DARK TO GREENISH-GRAY, MOD. SOFT TO MOD. HARD, OFTEN SOFT TO MOD. SOFT IN UPPER 4' ±, VERY SILTY CARBO. IN PART, SUB-FISSILE TO BLOCKY SLAKES ON EXPOSURE IN PART.
	COAL W U.C. (NO. 3)	0.4' - 1.0' ±	BLACK, HARD, SHALY, W/MED. GRAY, MOD. SOFT, SILTY U.C.
	SHALE (NO. 3)	40' ±	MEDIUM TO DARK GRAY, MOD. HARD, SANDY W/NUMEROUS TO OCC. SS. PARTINGS; GENERALLY SUB-FISSILE TO MEDIUM BEDDED.
	COAL W CARBO. SHALE	2' ±	BLACK, HARD, SLATY
	LIMESTONE	0.5' ±	GRAY, HARD, SHALY, FOSS.
	INDURATED CLAY	3' ±	GRAY, CALCAREOUS, SLICKENSIDED (UNDER CLAY).
	SILTSTONE	5' ±	GRAY, HARD.
	SHALE	7' ±	GRAY TO BROWN, MOD. SOFT TO HARD, FLAKEY, SLICKENSIDED.

PENNSYLVANIAN SERIES
LISMAN FORMATION

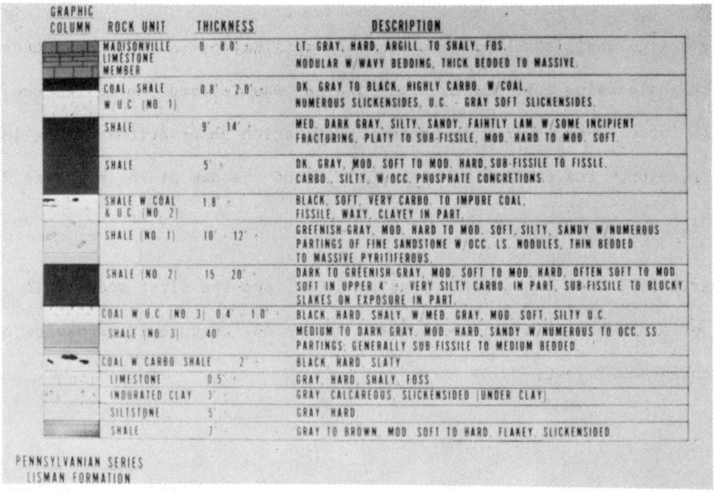

FIGURE 11. Description of Foundation Rocks at Uniontown Locks and Dam

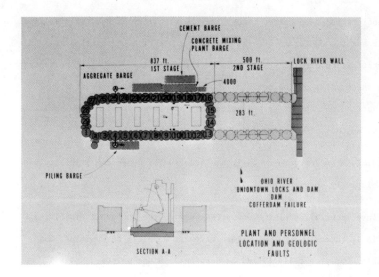

FIGURE 12.   Uniontown Locks and Dam Cofferdam
             Layout

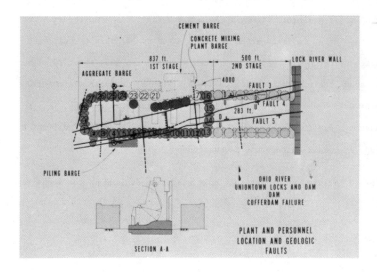

FIGURE 13.   Uniontown Locks and Dam Cofferdam after
             Februrary 1971 Failure

Questions had been raised concerning the stability of the cellular fixed weir and the cofferdam and the Louisville District had started an investigation program to re-evaluate the stability of these structures. The sheet piles of the cofferdam had been driven into rock at about elevation 295; the river was quite high at elevation 352. The normal water surface is at elevation 320.

At 9:40 AM on 26 February 1971, the question of cofferdam stability was conclusively answered.

Failure occurred under the downstream arm of the cofferdam, involving cells 17 through 23. The failure movement was like the opening of a double door with cells 16 and 24 serving as hinge points. Cells 17 through 21 formed one door; cells 22 and 23 formed the other. Both cells 17 and 22 were destroyed; the other cells translated to their final positions with minor tilting. Figure 13 is a diagram of the cofferdam layout after failure. Figure 14 is a photograph of the cofferdam after failure. Prior to failure the cells appeared to be draining well with seepage exiting from weep holes placed at elevation 312, about 10 - 15 feet above the excavation floor. During failure, no water was observed coming from beneath the cells; all the water came from the top. A pressure ridge formed between the moving cells and a normal fault which trends diagonally across the foundation.

Prior to failure, an exploratory hole was being drilled in the downstream

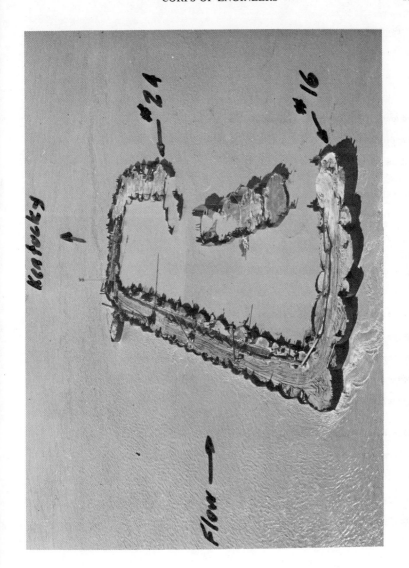

FIGURE 14. Photograph of Uniontown Cofferdam after Failure.

portion of the pier 7 foundation. The night before failure the hole
had been advanced through coal No. 2 at a depth of 9 feet. The morning
of the failure the hole was making a small amount of water. The driller
was standing by waiting for a core barrel. He heard water running down
and out of the hole. He looked into the hole and saw that it was
displaced, apparently below the coal. At that moment the alarm sounded
and he departed.

The failure apparently occurred in the coal No. 2 and underclay zone,
allowing translational movement of the cells and the top 10 to 15 feet of
rock as a unit. Figure 15 is a geologic profile through pier 8.

The cofferdam was rebuilt. All the debris and disturbed rock was removed,
the cells were reconstructed and filled. Before pumping the work area
dry, pressure relief holes were drilled all around the inside periphery of
the cofferdam. Twenty-one piezometers were placed in the cofferdam to
monitor uplift pressures in the No. 3 coal and under clay zone.
Provision was made to flood the work area when the river reached
elevation 335.

Additional explorations and analyses of foundation stability were
performed for the spillway pier foundations which confirmed the need for
founding within shale No. 3. However, founding in shale No. 3 would
require deep excavations at the Kentucky end of the cofferdam. A
construction sequence of partial excavations, concrete backfilling,

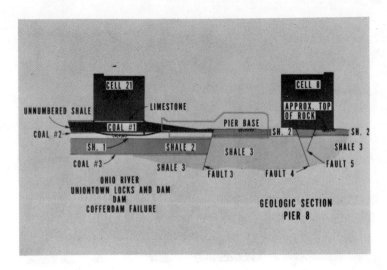

FIGURE 15.   Geologic Profile Through Uniontown Dam
             Pier 8.

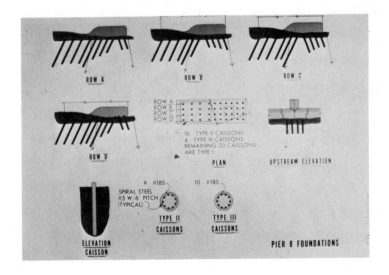

FIGURE 16.   Typical Caisson Layout for Uniontown Dam
             Pier Foundations.

FIGURE 17. Foundation Caisson Construction Activity,
Uniontown Dam.

FIGURE 18. View of a Nearly Completed Caisson Foundation
for a Spillway Pier, Uniontown Dam.

extensive bracing of excavations and more foundation drainage measures
would have been required to assure cofferdam stability.

After considering a number of alternatives it was decided to found the
piers on battered caissons with their tips within shale No. 3.

Figure 16 shows a typical caisson layout used for piers 6-11.

Data on the caisson foundations are as follows:

Number/Pier	40-45
Size	36" Diameter
Length	Up to 45 feet
Mabearing press	100 TSF (SH #3)
Bearing press used	82-92 TSF (Const loading)
Reinforcing	1-2 percent
Conc Str	5,000 psi

Figures 17 and 18 are views of caisson installation. The caissons were
installed and are performing satisfactorily.

In order to complete the project in the high water season, a 2-foot thick
concrete structure was placed between the cofferdam cells and the just
completed bases of the piers and sills. This procedure provided a brace
for the complete cofferdam inclosure. The cofferdam later withstood
river levels to its top.

CONSTRUCTION OF A PUMPED STORAGE POWER PLANT

ON SOIL FOUNDATIONS

by

Robert L. Polvi*

INTRODUCTION

This paper describes the construction of the power plant structures for the Ludington Pumped Storage Project located in Western
Michigan. It includes a brief description of the project, a more
detailed description of the power plant structures, foundation conditions, foundation preparation and treatment, problems encountered
during construction and the performance of the completed structures.

Ludington is quite a unique hydroelectric project in that there
were no tunnels, rock excavations or grouting required for the project.
All of the power plant structures are founded on sand and clay type
materials.

PROJECT DESCRIPTION

The Ludington Pumped Storage Project, owned by Consumers Power
Company and the Detroit Edison Company, is located on the east shoreline of Lake Michigan, approximately 55 miles north of Muskegon and
5 miles south of Ludington. The project is located on one of the
highest bluffs on the lake's eastern shoreline. The site encompasses
about 2000 acres and consists of a man-made upper reservoir, an intake
structure, six penstocks, a powerhouse with six reversible pump turbines,
a switchyard and a tailrace that includes two jetties and a breakwater.
(See Figure 1.)

The project is the world'a largest pumped storage installation in
service. It has a generating capability of 1,872,000 kilowatts. Each

---

*Construction Manager, Bechtel Inc., San Francisco, California.
Formerly Project Superintendent, Ebasco Engineering Corp.,
Ludington, Michigan

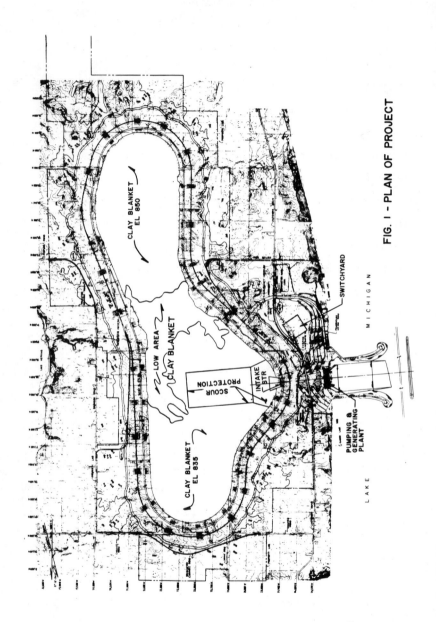

FIG. I - PLAN OF PROJECT

of the six units is rated at 312,000 kilowatts. The upper reservoir
has a capacity of 82,300 acre feet and is enclosed by an approximately
6 mile long, 39 million cubic yard, multi-zoned earth embankment, having
an average height of 108 feet, a maximum height of 170 feet and side
slopes of 2.5 horizontal to 1. The materials for the major zones in
the embankment were obtained essentially by selective excavations within
the reservoir, thus avoiding processing of materials. The reservoir
lining consists of a clay blanket varying in thickness from 3 to 5 feet
over the entire reservoir bottom, and from 8 to 10 feet thick on the
lower portions of the embankment. The interior face of the embankment
within the reservoir operating range is lined with hydraulic asphalt
concrete. The 28½-inch thick asphalt lining is composed of a 3-inch
hydraulic asphalt concrete subbase (HAC) layer, an 18-inch drainage
layer, a 2½-inch binder course and then covered with 2 - 2½-inch
hydraulic asphalt concrete layers with mastic seal coat. A seepage
monitoring system and dewatering system are installed in the drainage
layer.

The average water surface elevation in Lake Michigan is 580. The
maximum upper reservoir water elevation is 942. Thus, the static head
at maximum pool is 362 feet with normal drawdown of 67 feet.

The six unit intake structure located within the reservoir at the
nearest point to Lake Michigan consists of an intake apron with retaining
walls and splitter walls, a baffle wall to improve flow conditions and
six fixed-wheel gates with individual hoists. A 125-ton capacity gantry
is located on top of the structure for servicing the intake gates, hoists
and bulkhead gate. Six steel penstocks tapering in diameter from 28 feet,
6 inches to 24 feet, connect the intake structure to the powerhouse. The
upper end of each penstock is concrete encased where it passes under the
embankment. The incline sections on the 3 horizontal to 1 vertical penstock slope are backfilled with sandy material. The lower elbows adjacent
to the powerhouse are concrete encased. (See Figure 2.)

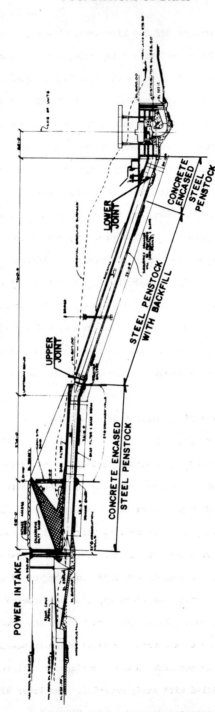

FIG. 2. – PROFILE ALONG WATERWAYS

The powerhouse is a semi-outdoor type, containing six pump turbines
and motor generators and all auxiliary mechanical and electrical equipment. Also included is a service bay and an equipment erection slab.
The 360-ton overhead gantry crane traverses the entire powerhouse,
service bay and erection slab. The main power transformers are located
on the upstream side of the powerhouse with 345 kV transmission lines
leading to the switchyard located at the top of the bluff south of the
penstocks. A conventional 5-bay, 345 kV switchyard, including all necessary switching and protection equipment, is provided on the site. Five
345 kV transmission lines lead from the project site to the Consumers
Power - Detroit Edison Distribution System throughout the state.

Two jetties extending approximately 1600 feet into the lake were
constructed on each side of the tailrace to protect the powerhouse and
tailrace channel from the erosive effects of Lake Michigan and the discharge velocities of the units. A breakwater approximately 1700 feet
long protects the entrance to the tailrace channel. It is located
approximately 2700 feet from the shoreline. (See Figure 3.)

Figure 4 contains a summary of characteristics of the major features
and the major subcontractors on the Project.

DESCRIPTION OF FOUNDATION CONDITIONS

The site geology and topography are the result of glacial and post-
glacial processes that began as late as 12,000 years ago during the last
ice advance. The high land on which the upper reservoir is located is an
upland morainic tract of glacial drift underlain by a succession of discontinuous clay till and sand outwash deposits. Bedrock formations are

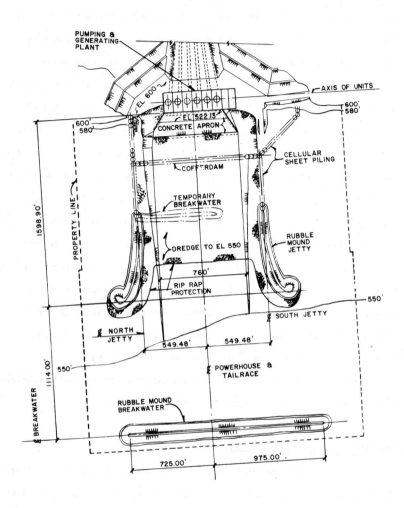

FIG. 3.- LAKE FRONT STRUCTURES

LUDINGTON PUMPED STORAGE PROJECT
FEATURE SUMMARY SHEET

OWNER: Consumers Power Company
The Detroit Edison Company

ENGINEER CONSTRUCTOR: Ebasco Engineering Corp. General Project Management, Intake, Penstock Encasement, Powerhouse, Tailrace & Auxiliary Mechanical Construction

MAJOR SUBCONTRACTORS:
Embankment & Clay Lining — Walsh-Canonie Companies
Asphalt Lining — Morrison-Knudsen-Stratbag
Penstocks & Spiral Cases — Chicago Bridge & Iron Co.
Pump-Turbine, Motor-Generators — Hitachi America, Ltd.
— Wismer & Becker
Power Plant Electrical — Newkirk Electric Associates
Cofferdam — Canonie-Buitema Companies
Jetties & Breakwater — Great Lakes-Canonie, A Joint Venture
Tailrace Excavation — Luedtke Engineering Company
Switchyard Construction — M. J. Electric, Inc.

Peak Work Force — 2,850 Men

Total Installed Capacity — 1,872 MW — 6 Units @ 312 MW

EMBANKMENT
Type — Asphalt lined earth fill
Volume — 39.3 million cubic yards
Elevation top of dike — 950'
Elevation toe of dike — 880–780 (varies)
Average height — 108'
Maximum height — 170'
Asphalt lining — 29,200 lin. ft., — 520,000 tons

RESERVOIR
Lining material — 5' clay blanket — 6.35 million c.y.
Volume — 82,300 acre feet
Bottom Elevation — 830–845' (varies)
High water elevation — 942'
Low water elevation — 875'
Draw down — 67'

POWER INTAKE
Bottom Elevation — 805'
Top Elevation — 950'
Concrete — 58,000 c.y.
Reinforcing steel — 5,461,000 pounds
Crane — 100 ton gantry
Gates — 6–29' x 29' fixed wheeled, with 250 ton hoists

PENSTOCKS
Number & shape — 6 circular
Material — Steel — 17,000 tons
Encasement — Concrete at embankment & powerhouse — 158,000 c.y. Concrete saddle and earth backfill on slope — 225,000 c.y. random fill
Diameter — 28'6" taper to 24'
Length — 1335' each
Area (ave.) — 490.9 sq. feet each
Velocity (ave.) — 24.c ft./sec.
Discharge Capacity — Total — 73,500 cfs

POWERHOUSE
Type — Semi-outdoor
Dimension — 576' x 170' x 106' high
Excavation — 700,000 c.y.
Concrete — 186,000 c.y.
Reinforcing Steel — 20,000,000 pounds
Forms — 1,000,000 SF
Crane — 360 ton gantry

Pump/Turbines
Manufacturer — Hitachi
Type — Frances - Reversible
H.P. — Pump — 433,000 hp
Elevation & Distributor — Turbine — 413,000 hp @112.5 rmp
Max. Head — 553.5'
Min. Head — Gen. Mode 353' -- pump mode 373.5'
Gen. Mode 286'' -- pump mode 305.0'

Generator-Motor
Manufacturer — Hitachi
Type — Outdoor synchronous 3 phase, 60 cycle, 20kv, 945 amp.
Rating — 112.5 rpm
Capacity — each — Gen. mode 310 mva @ 60 and .85 power factor Pump mode 337 mva @ .85 power factor and 80°C

SWITCHYARD
Type — 345 KV — 5 bays
Transformers — 10 ASEA 20KV/345KV single phase
Air blast reversing switches — 40 Brown-Boveri 20KV
Air blast circuit breakers — 13 Cogenel 345 KV

TAILRACE
Jetty sheet pile cells — 1820'
Jetty rubble mound — 1480'
Breakwater — rubble mound — 1700'
Excavation — 1,100,000 c.y.
Concrete apron & retaining walls — 28,000 c.y.
Area (average between jetties) — 31,530 sq. ft.
Velocity (average between jetties) — Gen. mode 2.3 fps
W/6 units in operation) Pump mode 2.1 fps

FIG.4.-FEATURE SUMMARY

at a depth of 800 feet. Because of the complex stratigraphy and the
critical nature of the predominantly sandy soils underlying the project
area, extensive subsurface exploration programs were performed. Initial
studies began in 1961 and were intermittently continued through 1970 when
final design of the present reservoir and power plant structures were
essentially completed.

The soil conditions encountered in the foundations of the power
plant structures were generally as follows: The intake apron, intake
structure and upper penstock encasement were generally founded on dense
site sand and calcareous silty sand. The 3:1 penstock slope consisted
primarily of sandy materials with three undulating clay layers at elevations 610, 650 and 710. There was perched water emerging from the
upper surface of each of the clay layers. The lower penstock encasements were founded partially on sand and clay. The entire powerhouse
and tailrace slab were founded on clay. A description of the general
types of material, i.e. site sand, calcareous silty sand and clay, is
as follows:

Site Sand - A medium- to fine-grained sand with an average fines
content of less than 10%.

Calcareous Silty Sand - A cohesionless sand having a fines content
between 10% and 35%; it is reactive to dilute hydrochloric acid and was
identified during construction by its pink to red color when moist.

Clay - A silty clay having a plasticity index of approximately 15
and a liquid limit of 30.

DESCRIPTION OF STRUCTURES

Since the structures were founded on sand and clay, the design considerations included consolidation of the foundation and differential settlement of the structures. The structures contained heavy reinforcing in the foundation lifts. The intake and upper penstock encasements were designed as three - 2-unit monoliths with continuous reinforcing across each monolith. The contraction joints were located between Units 2-3 and 4-5. The contraction joints contained shear keys and waterstops. In the powerhouse area, each unit was designed as an individual monolith. Contraction joints, including shear keys and waterstops, were located between each unit. A brief description of each of the power plant structures is included below.

Power Intake. The power intake structure consists of an intake apron, retaining and splitter walls and the main gate structure. The intake apron extends 293 feet upstream of the gate structure and expands in width from approximately 231 feet at the gate structure to 500 feet at the upstream edge. The splitter walls are located between Units 2-3 and between Units 4-5 and are designed to improve flow characteristics primarily in the pumping mode. Walls located on each side of the intake apron extended 193 feet upstream of the gate structure and retain the upstream dike slope from the intake opening. An 8-foot (minimum) thick clay blanket was placed under the concrete intake apron slab. A 20 mil PVC membrane was placed between the clay blanket and the apron slab to provide an additional seepage barrier. A rubber (EPDM) membrane waterstop is embedded in the concrete and extends a few feet into the clay to form a water-tight intersection between the clay lining and the concrete retaining walls, apron and intake structure.

The intake structure is 65.feet upstream/downstream, 231 feet wide
and 145 feet high. It contains 6 individual penstock openings, a fixed
wheel gate for each opening and an individual gate hoist for each gate.
Bulkhead gate slots also are included for each penstock opening. One
bulkhead gate is provided for servicing the intake gates. The gantry
crane also is designed to handle the bulkhead gate. A baffle wall for
elimination of vortices at the intake entrance is located approximately
30 feet upstream of the structure and rests on the top of the splitter
and retaining walls, 50 feet above the floor of the apron. The baffle
wall is a prestressed, precast structure with 4-foot diameter perforations
spaced 7 feet vertically and from 14 to 24 feet horizontally. A cross-
section of the intake structure is shown on Figure 5.

Penstocks. There are 6 separate penstocks, varying in diameter
from 28.5 to 24 feet and approximately 1335 feet long. The upper section
located under the dike is encased in concrete. The maximum height of
embankment over the encasements is approximately 100 feet. The upper
penstock encasement extends from the intake structure a distance of
approximately 530 feet to the downstream toe of the embankment. The
encasement provides a 5-foot thick reinforced concrete cover around the
penstocks between the intake and the centerline of the dike and a 4-
foot thickness from the centerline of the dike to the downstream end.
A drainage gallery is provided under the penstock encasement at the
centerline of the dike. Two access galleries lead from the downstream
end of the encasement to the drainage gallery.

The penstock steel was designed to withstand the internal hydrostatic pressure and the reinforced concrete encasement has been designed

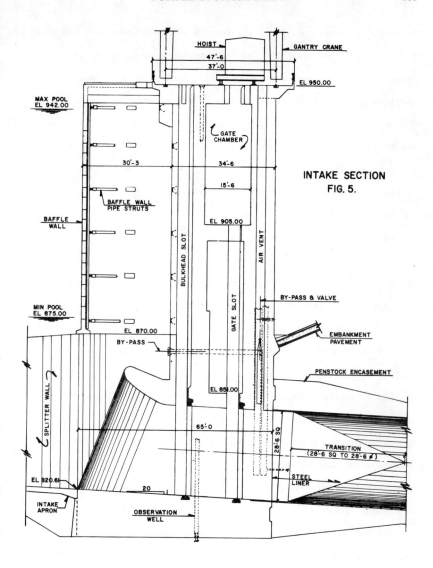

INTAKE SECTION
FIG. 5.

to take the embankment load. The section of penstock between the downstream toe of the embankment and the powerhouse is supported on a continuous concrete saddle and is completely backfilled with sand. The
sections of penstock that are backfilled are provided with stiffeners at
5-foot intervals to withstand the external earth pressures. To provide
for thermal expansion and contraction and differential settlement between
the encased and buried penstock sections, two differential settlement
joints are provided, one at each end of the backfilled section. These
joints also provide an allowance for creep of the slope without damage
to the penstocks. The articulated joints are capable of taking 4 inches
of longitudinal expansion or contraction and 2 inches of differential
settlement, while maintaining their integrity. Each joint consists of
a 21 foot 3 inch long sleeve with packing on upstream and downstream
ends. The lower joints are enclosed in individual chambers that are
accessible through manholes. The upper joints are exposed and are
easily accessible via walkways.

An extensive drainage system protects the penstock slope from
underseepage. The system includes a gravel drainage layer with collecting
fiberglas pipes. Horizontal drains, installed during construction to
intercept the natural spring flows, have been incorporated into the
permanent drainage system. The drainage system includes provisions for
monitoring flows in the drains. Penstock retaining walls are provided
on each side of the penstocks at the downstream toe of the dike. The
penstock encasement below the lower differential settlement chamber,
a distance of approximately 40 feet, is heavily reinforced and tied to
the powerhouse structure. A 4-inch contraction joint is provided in
the differential settlement chamber to separate buried penstock and

concrete chamber from the lower encasement that is tied to the powerhouse. This space also provides a means of monitoring movement of the penstock on the slope.

Powerhouse. The powerhouse is a semi-outdoor type and is conventional in design except that it is bearing entirely on clay and therefore requires a heavily reinforced substructure. It consists of six 85-foot by 170-foot bays from upstream to downstream faces. The structure is approximately 105 feet high. An erection bay, including a machine shop, is also provided for servicing the units. A 360-ton overhead gantry crane is provided on the roof deck. The main power transformers are located on backfill upstream of the powerhouse and generally between the penstocks. The main gantry crane is sized to handle the pump turbine runner that can be serviced through the rotor that has a removable spider. The generator rotors and stators are too large to be handled in one piece by the gantry crane. Each stator was delivered in six sections and erected in place. Each rotor was completely assembled in place, i.e., spider assembled, rim steel stacked and poles mounted.

A reinforced concrete slab is provided in the tailrace extending out 193 feet from the downstream face of the powerhouse. Retaining walls are provided on each side of the apron slab for retaining of powerhouse backfill. The slab diverges to a width of approximately 741 feet and slopes up to elevation 550 (30-foot water depth) at the downstream edge. Downstream of the apron the tailrace bottom is lined with riprap for a distance of 200 feet and then with a 1-foot thick layer of cobblestone for an additional 900 feet. A combination of sheet pile cells and rubble mound rock forms jetties on the north and south sides of the tailrace channel. A rubble mound rock breakwater is located at the entrance to the tailrace, as shown in Figure 3.

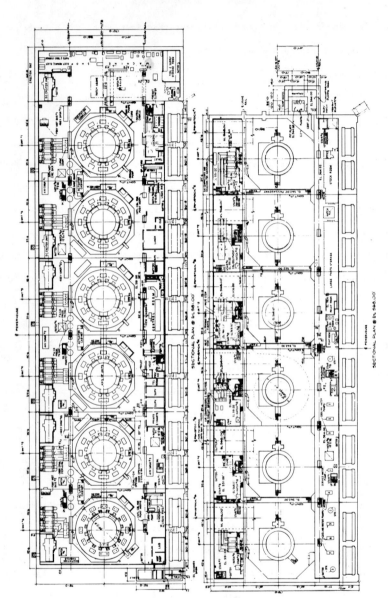

FIG. 6. – POWERHOUSE PLANS

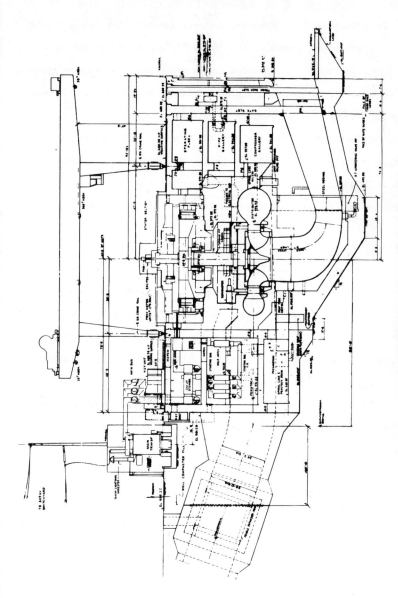

FIG. 7. – POWERHOUSE CROSS SECTION

The powerhouse excavation extends approximately 85 feet below lake level. A wellpoint eductor system, installed around the periphery of the powerhouse excavation provided a very satisfactory unwatering system. The wellpoints were spaced at approximately 10-foot centers in the more pervious sand and 20-foot centers in less pervious zones. Most of the wellpoints extended through the sand and down to the top of the major clay layer. The total discharge from the wellpoints system ranged from 3000 to 2000 gallons per minute throughout the construction period. A large portion of this water was used for construction purposes.

FOUNDATION TREATMENT AND CONSTRUCTION METHODS

The foundation excavations were performed generally with heavy equipment. The intake foundation, penstock slope and powerhouse foundation were excavated with scrapers and bulldozers. Slope trimming and removal of material was performed with small bulldozers, front end loaders and dump trucks. Due to the effective wellpoint system around the powerhouse and the horizontal drains that penetrated up to 300 feet into the penstock slope, water seepage was not a serious problem during excavation. The most serious water problem occurred in September, 1970, when a severe storm with winds up to 90 mph blew the fifty-foot high plumes of breaking waves over the cofferdam into the powerhouse excavation causing serious erosion of the berms on the inside of the cofferdam and filling the excavation with about 20 feet of water. The berms and drainage system were enlarged and improved to protect against future storm damage.

The foundations were excavated as close as possible to the desired grade and then final trimming was performed by hand. A 3- to 4-inch layer of lean concrete (mud mat) was placed over the foundation material immediately after completion of the excavation. Prior to installation of the mud mat, the foundation was checked by probing to assure that soft areas were not present. In localized cases, it was necessary to remove additional material and replace it with compacted sand. The only area where special foundation treatment was performed under the mud mat was under the upper penstock encasement between the center-line of the dike and the downstream toe. In this area, a 1-foot 3-inch filter sand layer was placed on top of the foundation material and then a 1-foot gravel drain layer covered by a 20 mil PVC lining was installed. This provided a drainage layer under the penstock encasement downstream of the transverse drainage gallery. In most instances the mud mat was placed immediately after the foundation material was excavated to grade. The mud mat was required to maintain the shape of the foundation during construction and to provide a base for anchoring forms and supporting reinforcing steel for the initial lifts of the structures. The mud mat concrete very effectively maintained the foundation shape in the sand zones and protected the clay zones from drying and shrinkage, thus reducing structural concrete overruns.

Conventional methods were used for construction of the concrete structures. Truck-mounted and crawler cranes were used for placing forms, reinforcing steel and other embedded items. Concrete was placed with 4 cubic yard buckets handled by crawler cranes, supplemented by a creater crane and concrete pumps. A trestle was constructed across the electrical bay on the upstream side of the powerhouse approximately 10 feet above the top deck. Two revolver cranes were utilized on the trestle for supporting the work crews in preparation of concrete pours and for concrete placement. During the peak construction period in 1971, a maximum of 33,000 cubic yards of structural concrete was placed in a month. The total concrete required for the project structures exceeded 440,000 cubic yards. A peak work force for the entire project of approximately 2850 men was reached in mid-1971.

PROBLEMS DURING CONSTRUCTION

One of the major problems encountered during construction was maintenance of excavated slopes. The sandy slopes being very erodible received extensive damage during the frequent torrential rainstorms that occurred in the early summer and fall seasons. The most serious erosion experienced during the construction period was on the penstock slopes and in an adjacent natural draw called Kings Canyon. During the period when the penstock slope was being excavated, erosion control structures could not be maintained due to heavy equipment traffic. After bulk excavation was completed, temporary drainage dikes, ditches and other structures were provided to divert water away from the top of the excavated slopes and to carry the runoff across the 3:1 and 2:1 slopes. Erosion control on the penstock slopes between the intake and powerhouse continued to be difficult and required constant maintenance due to the continuous construction activity. The dike construction adjacent to the intake and the construction facilities around the intake and upper penstocks enlarged the drainage area contributing to runoff toward Kings Canyon and the penstock slope.

To reduce runoff several percolation ponds were constructed in the drainage area above the crest of the slope. A 36-inch diameter steel pipe was installed in Kings Canyon and 18-inch diameter CMP pipes were installed on each side of the penstock slope to carry water down the slopes to Lake Michigan. All runoff at the top of the slopes was diverted toward these pipes. Transverse ditches were constructed on the penstock slope at 50-foot intervals in elevation to carry water to the 18-inch pipes. Erosion gullies were filled with hand compacted crushed stone. The ground water seepage in Kings Canyon was controlled by utilizing horizontal drains, a sand gravel filter-drainage zone and 6-inch diameter PVC slotted pipes leading to permanent drainage pipes at the bottom of the slope. The entire canyon then was backfilled and reshaped to eliminate the natural gulley.

On the penstock slope (3:1), several methods of erosion control were used. A combination of burlap and gravel cover proved to be the most effective during construction. The slopes were seeded and mulched as soon as possible after completing final trimming and construction of the permanent drainage structures. Before the vegetation cover developed, many of the permanent drainage pipes and catch basins were filled with sand from slope erosion and had to be cleaned out. It was finally proven that the most effective method of stabilizing the excavated slopes was to develop a good vegetation cover with adequate temporary and permanent drainage facilities to handle the runoff.

Another problem encountered during construction was compaction of sand backfill under the penstock sections. The design required that sand be compacted to a relative density of 75%. With a test section several attempts to compact the material under the penstock sections using various types of compaction equipments and methods proved to be unsatisfactory. It was finally decided that although the sand

probably could be compacted, it could not be tested to verify that
densities met the specification requirements within the 12 inches to
18 inches immediately under the penstocks. Therefore, it was decided
to replace the compacted sand backfill with a 120° lean concrete saddle.
This saddle was placed under the pipe in 90-foot sections utilizing
concrete pumps. Erosion gullies mentioned above developed under some
of the penstocks to a depth of 4 to 5 feet. Hand backfilling methods
were not satisfactory so all loose material was removed and lean concrete was used for backfill in conjunction with the saddle construction.
After completion of the concrete saddles, an 18-inch gravel drainage
layer including 6-inch diameter slotted PVC pipes was placed on the
slope between the penstocks. Filter was placed on the top and bottom
of the drainage layer. Sand backfill was then placed in 6-inch
horizontal lifts and compacted with vibratory smooth drum rollers.
Hand placement and compaction was necessary adjacent to the penstocks
and around the stiffener rings. After testing various equipment loadings
on the penstocks and measuring the stress in the steel, it was approved
to pass the equipment used for the backfilling operation over the
penstocks providing a minimum 2-foot cover was maintained. Backfill
was placed to a depth of 4 feet over the top of the penstocks.

PERFORMANCE OF STRUCTURES

        Initial pumping with Unit No. 1 was performed on October 23, 1972.
The reservoir was filled at a slow rate during the following month. The
first unit was placed in commercial operation on January 17, 1973. The
following units were placed in commercial operation at approximately 2

month intervals with the last unit in commercial operation on October 1,
1973. So the major structures have been in service for over a year. As
of this date all power plant structures appear to be performing as
designed. There is an extensive instrumentation and monitoring system
for the various structures. The performance of the structures to date
is described below.

Intake. Slightly over 1-3/4 inches of settlement has been measured
on the intake since completion of the structure. Most of the settlement occurred during initial filling of the upper reservoir and is
believed to be related primarily to the water load in the general area.
The settlement measurements indicate that the structure settles and
rebounds about 1/2 inch each day in response to maximum-minimum pool
levels. There has been no transverse or longitudinal movement of the
structure. Also, there is no noticeable differential settlement at the
contraction joints.

Measurements in the five observation wells at the intake revealed
a rise in ground water table during initial reservoir filling. The
rise was so closely related to reservoir fluctuations that it was
believed to be caused from leakage through the concrete apron, retaining
wall or intake construction joints. The reservoir was lowered and all
joints suspected of leakage were chemically grouted. This proved to
be partially effective. The initial ground water table was about
40 feet below the intake foundation. It raised to about the elevation
of the bottom of the intake and then began decreasing. It has been
decreasing for the last year and now is about 20 feet below the bottom
of the structure.

Penstocks. The upper penstock encasement has settled about 1 inch
at the centerline of the dike, measured in the transverse gallery, and
about 3/4 inch at the downstream end near the toe of the dike. The
penstock retaining walls have settled less than 1/2 inch. These settlements do not appear to be affected by reservoir fluctuations. There is
no noticeable differential settlement at the contraction joints. Observation and relief well measurements in the transverse gallery and at the
downstream end of the penstock encasement show a water table rise but
the levels are lower than at the intake and well below the concrete encasement, thus indicating a gradient away from the intake structure.

The penstocks on the 3:1 slope have settled a total of slightly
less than 1/2 inch since initial filling. There has been no measurable
creep down the slope. Dial gages and extensiometers at the differential
settlement joints measure movements of the penstocks that have been
correlated directly to ambient air temperature and water temperature
changes. These movements vary from 1/2 inch to 1 inch from season to
season. Leakage through the differential settlement joints is essentially
stopped by packing adjustments after initial penstock filling. There is
no other penstock leakage. Strain gages provided to monitor the effect
of backfill loads and operating stresses indicate maximum actual stresses
are well within the design limit.

Powerhouse. The powerhouse structure has been completed for
nearly 2 years. The total maximum measured settlement is approximately
1/8 inch, and is considerably less than the design values. Future
settlement is expected to be negligible. Control and expansion joints
in the structure are performing as designed with no detectable differential settlement. Leakage through the powerhouse walls and the

contraction joints does not exceed 5 gpm. There has been no other detectable movement of the powerhouse structure. The vibration from the pump turbine - motor generators is very small and has had no measurable effect on settlement or movement of the powerhouse structure. The ground water table around the powerhouse has stabilized at approximately the same elevation as Lake Michigan or 20 feet below the top deck. This was considered in the design and has caused no problems.

In conclusion, it is demonstrated that major power plant structures that are normally founded on rock can successfully be designed and constructed on dense soil foundations without enduring great additional difficulties and expense. In my opinion, the foundation conditions had no significant effect on the overall construction schedule.

JOSE Mª de ORIOL DAM

(ALCANTARA)

by Manuel Castillo
Engineer de Caminos
Hidroelectrica Espanola S.A.
Madrid, Spain

This dam lies on the Spanish part of the Tagus River not far from the Portuguese frontier, in the municipal district of Alcantara, Caceres Province.

The name, "Alcantara", is of Arabic origin, meaning "the bridge". The bridge referred to here is a famous one built in the days of the Roman Empire (which included Spain), completed early in the second century of the Christian Era, during the reign of Emperor Trajan, who himself happened to be also Spanish by birth.

The bridge has been admired by all those who have seen it over the ages. Hence the Arabic word, surpassing its original meaning, was given to the adjacent town and its district.

As far as situation was concerned, the building of a dam on the Tagus River at this point of its course was subject to two governing factors. On the one hand it could not be placed downstream from the Roman bridge because the latter's archaeological value as an outstanding monument meant that interference with it was quite out of the question. Upstream, on the other hand, just over a mile from the Roman bridge, there is another limiting point, where the Alagon River flows into the Tagus. The Alagon is one of the most important tributaries, providing about 30 per cent of the water carried by the Tagus. Thus, within this stretch of 1,800 metres, only one section existed with suitable topographical features for the building of a dam of such a height. The situation thus decided nevertheless, raised

certain geological problems which explain the extent of the preliminary surveys and also the unique feature of this work, where the results achieved are judged to be highly satisfactory as regards both the technical and economic aspects.

Several projects were drafted for damming the Tagus in the neighbouhood of Alcantara, with the dam always placed at the point already mentioned and with the power station at its foot. Several structural forms and heights were envisaged, considering that the resultant reservoir was bound, above a certain reservoir level, to affect an important highway and railway communications knot 50 kms upstream from the dam. The solution adopted in the end was the most ambitious of them all, both in terms of height, taking advantage of practically all the possibilities of the narrowing, and in terms of installed capacity. Hence, at the time of its inauguration, this proved to be the most important artificial lake in Spain, with the hydro-electric power station of highest capacity in Western Europe.

The criteria governing the terms of the project finally adopted, known as the 1965 Project, were as follows :

— The stage reached in 1965 as regards the tendency to specialize in hydroelectric installations for "peak" power production.

— Improved definition of flood flow and discharge.

— More complete information on the geology of site available since 1958.

The characteristics of the final project are given below.

Work on diverting the river began in 1964, at the same time as the surveys and studies of the foundation rocks in the form of tunnel tests and soundings continued.

What we call the preconsolidation work -one of the most interesting technical aspects of this work- lasted three and a half years, from January 1964 until July 1967.

Excavation of the dam started in September 1964 and was completed in November 1968. This excavation was obviously dependent upon progress in preconsolidation.

Concreting of the dam began in July 1966 and the last cubic metre of concrete was laid in May 1969.

The four power station sets were gradually brought into operation between November 1969 and July 1970.

The dam was officially and formally opened by the Head of the Spanish State on July 7th, 1970. Subsequently, on October 27th of the same year, the latter further honoured Don José María de Oriol, the Chairman of the Boards of Directors of Hidroelectrica Española, S.A., by naming the installation after him.

CHARACTERISTICS

Of the Watershed

Surface	51,916 sq km
Average annual rainfall . .	680 mm
Average annual flow	7,691 x 10^6 cub m
Maximum known annual flow (1963)	17,920 x 10^6 cub m
Minimum known annual flow (1950)	2,406 x 10^6 cub m
Maximum known annual flood (1941)	11,104 cub m/sec
Minimum known flow (1950)	5 cub m/sec

Of the Reservoir

Total volume	3,137 x 10^6 cub m
Serviceable volume	2,408 x 10^6 cub m
Maximum level (with excess rise)	22 m a.s.l.
Maximum exploitable level	218 m a.s.l.
Minimum exploitable level	180 m a.s.l.

Of the Dam

Type : Eased gravity with double abutment.

Maximum height above foundations	130 m
Maximum height above bed	123 m
Length of crest	570 m

Dam spillway :

Maximum flow	4,000 cub m/sec
N^o of waterways	3

Lateral spillway :

Maximum flow	8,000 cub m/sec
N^o of waterways	4

Dewatering conduits :

Maximum flow	2 x 300 cub m/sec
N^o of drains	2

Total maximum drainage capacity :

Via spillways and dewatering conduits	12,600 cub m/sec
Via machinery etc.	1,172 cub m/sec
Total	13,772 cub m/sec

Of the Power Station

Turbine capacity	915,200 kw
Pumping capacity	—— kw
Maximum gross head	108 m
Minimum gross head	70 m
Computation head	99.10 m
Average annual production	1,800 x 10^6 kwh

PRELIMINARY STUDIES

Hydrology of the Tagus River

At Alcantara, the Tagus River collects the inflow of a watershed of 52 sq kms area. Under natural circumstances, the latter is typified by an extremely irregular hydrographical system which, over the last thirty years, has registered from a low-water flow of less than 5 cubic metres per second, rising to peak floods of over 11,000 cub m per sec.

This same irregularity marks the geographical distribution of the in-flow, showing clear assymetry between the two banks : scanty in-flow on the left and abundant in-flow on the right bank, especially in the lower stretch which receives the run-off from the Gredos Mountains, a barrier which is well placed to draw from the Atlantic storms brought into the peninsular on south-westerly winds.

As from the late nineteen-forties, when surveys were begun concerning the integral exploitation of the lower Tagus, there has been constant interest in improving the information on its hydrological system. Gagings were carried out at the most representative points, limnigraphic stations were set up, the meteorological characteristics of

the zone were studied, especially in terms of the rain-bearing south-westerly winds, and as many statistics as would contribute to providing a more complete picture were carefully collected. These activities have continued and been improved upon since the projects were completed.

During this same period, the Tagus basin in general, underwent far-reaching changes as a result of the construction of many works controlling the water-flow, directed at improving water-supply and irrigation schemes, all of which represent at the present time a storage capacity of more than 10,000 million cubic metres, of which 50 per cent are covered by the works of Hidroelectrica Española alone.

This meant that when the Alcantara project was launched, adequate hydrological data were available both for the evaluation of hydroelectric productivity and for correctly working out the size of the dam and the evacuation arrangements. Standing norms governing Large Dam projects at that time envisaged the employment of probabilistic methods of establishing the size of the spillways, taking as the basis those flow rates with probabilities of occuring once in five hundred years. Thanks to the completed surveys, registered data were available for 62 "points" of annual flow, thus providing reasonable gua-

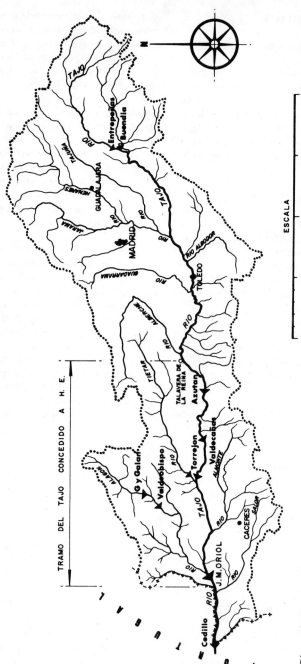

Situation map. Tagus river catchment area.

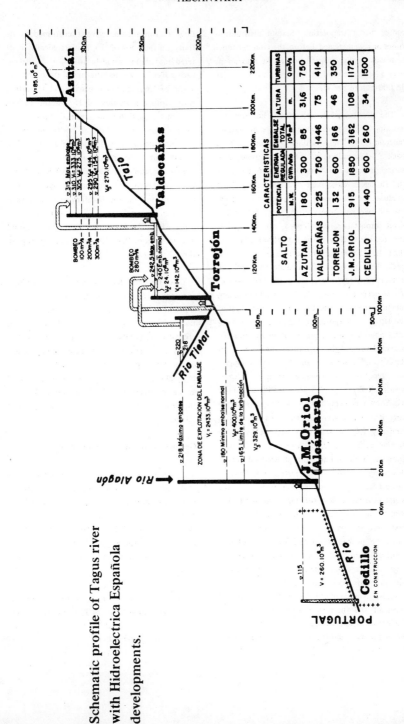

Schematic profile of Tagus river with Hidroelectrica Española developments.

SALTO	POTENCIA M.W.	ENERGIA REGULADA GWh/año	EMBALSE TOTAL 10⁶m³	ALTURA m.	TURBINAS Q m³/s
AZUTAN	180	300	85	31,6	750
VALDECAÑAS	225	750	1446	75	414
TORREJON	132	600	166	46	350
J.M.ORIOL	915	1850	3162	108	1172
CEDILLO	440	600	260	34	1500

rantee for extrapolation. Several distribution functions were tested before the Gumbel function was finally adopted, where the maximum over a return period of 500 years works out at 15,000 cub m per sec.

The problem was not solved, however, by working out the maximum foreseeable instantaneous flow. The Alcantara Dam was to hold back a 3,130 million-cubic-metre reservoir, measuring more than 100 million square metres at full storage level, capable of a laminating effect because the likelihood of a temporary excess rise of two or three metres, seemed to be of scant importance. It followed, obviously, that the maximum flow to be led off would be less than the 15,000 cub m/sec foreseen as in-flow peak. But, in order to calculate this and the levels reached in the reservoir during the outflow, it was necessary to obtain the wave hydrogram, remembering that the theoretical estimation of the latter for purposes of reconstructing the natural phenomenon of precipitation and run-off, provided only slight guarantees for such an extensive and complex basin. However, limnigraphic records were available for many streams with peak flows of as much as 7,500 cub m per sec. Shown as a graph, with their vertices coinciding, it could be seen that there was an obvious tendency for the waves to reach peaks as the maximum flow increased. In other words, flow duration over a certain figure showed a decrease, this being an extremely interesting consideration to be taken into account when forecasting the worst possible in-flow situation. This led to the application, for what we understand was the first time in Spain, of extrapolation methods based on probability distribution for volumes pouring in on in-flowing waves at the critical stage, this stage being defined as that marked by the 48 hours time span on both sides of the peak. Adjustment of the Gumbel function proved excellent and the results were contrasted with the linear correlation between the critical volume of the wave and the maximum instantaneous

flow. This meant that waves transformed by affinity on the basis of the limnigraphically registered maximum, could be used for calculating excess rise in level and the size of the outlets. A study of lamination was carried out for affine waves with a peak flow of 15,000 and 18,150 cub m/sec, and for an assumed limit of the homothetic wave at a ratio of 2 of the maximum registered double flow for double duration. The 15,000 cub m/sec maximum-flow affine wave, would dewater when all the mobile elements were employed, but without the hydroelectric power station and the dewatering conduits, with a maximum out-flow of 11,862 cub m/sec. In other words, lamination in the reservoir would lead to a more than 20 per cent drop in peak flow.

Likewise, available hydrological information enabled us to carry out studies to foresee diversion and cofferdam capacity during the construction, protection against possible floods during critical periods and the inclusion of hydroelectricity usage in the simulated exploitation models of the firm's power producing system. Estimated yield for the dam worked out at an average annual production of 1,900 million K.W.H. on the basis of the hydraulicity of the previous 20 years.

Geology of the foundation ground

In geological terms, the Alcantara gorge is uniformly Cambrian, consisting of sub-vertical, WNW-ESE pointing slatey shreds and schists. These schists, of Pelithic origin, having been more or less lapidified under the effects of regional metamorphism, lie in the form of thin strata, mingling with and locally crossed by very thin quartz veins.

The cracks in the bedrock reflect both its Hercynian formation as also orogenic movements of a later date, which have given rise to a highly complex structural pattern.

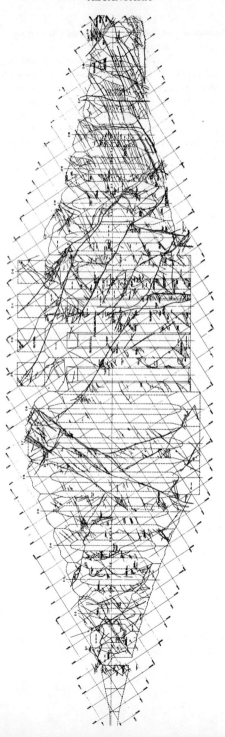

Permeability of the foundation and hillside bedrock.

Work carried out "in situ" prior to 1958, and that done subsequently with extraordinary detail and profusion, all goes to show that the bedrock as a whole, is homogeneous and that there is no fault or crack representing a general discontinuity in its geology. This is a rocky core which has remained preserved "in situ" since it was metamorphosed into schist, a core with deep roots and one which over its "long life" has undergone an extensive yet orderly process of fracturing which has partly made use of the active agency of meteorization.

This systematic, orderly fracturing of the upper parts of the dam-foundation bedrock spoke of difficulties not merely in terms of rock resistance, especially with regard to the geological behaviour of the foundations, but also in terms of difficulties due to permeability, which, whilst they were never envisaged as important as far as leakage from the reservoir was concerned, were always borne in mind when examining the negative effects of sub-pressure phenomena.

Such factors, in which the permeability of the foundation rock played a key role, led us to opt for the present type of dam as being the best suited to counter any problems arising from the sub-pressures which might well occur in such a type of foundation rock as this particular one.

It was for this same reason that we decided to apply a method involving the energetic and systematic treatment of the rock on the actual site, in an aim to return to it part of its qualities of cohesion, elasticity, compactness and impermeability, which had been diminishing on the surface zones in the course of several geological ages. In applying this treatment, we had the support of past experience in Spain, albeit locally and on a limited scale, in such cases as Valdecañas and Aldeadavila, to mention a few examples.

A description follows later giving details of this preliminary averall treatment of the rock the dam. It involved approximately one and a half million cubic metres of rock. Considered merely in terms of seepage, the results have been satisfactory. Even more satisfactory were the results obtained in terms of the transitional homogeneization and improvement of the elastic module underlying the system as envisaged and applied with great savings in time and final costs.

Returning to the question of permeability, while the reservoir was being filled, it was found that seepage into the drains downstream of the main screen amounted to a maximum of 173 l/min, i.e. about 3/litre sec.

The fact that wet patches arose on the foundation rocks in some sections of the dam, and that somewhat high sub-pressure was recorded by some of the piezometers, led us to introduce some additional drains. These got rid of the inconvenient wet patches and lowered the sub-pressure readings, which were above the normal readings for the rest of the dam, to a satisfactory level. Seepage into the drains obviously increased, until it reached stability at about 270 l/min for the stage of maximum storage (level 216.60), involving 300 drains along the line of contact between rock and concrete.

The reservoir : geology and permeability

The reservoir lies basically upon primary bedrock, consisting of slate formations, granits, quartzitic strikes, covered in some areas by quaternary deposits.

TREATMENT

As far as the impermeability of the basin was concerned, the granite, with its joints fothered by caolinization and thanks to its arrangement in the form of islands surrounded by impermeable metamorphic rocks, offered a fair guarantee of impermeability.

The Azoic or Cambrian slates within the basin, may be considered as absolutely water-tight. The same may be said of the Middle and Upper Silurian formations.

The armorican quartzite, which builds a long but not strong syncline -when compared with the overall dimensions of the watershed- is flooded at different points. As the distance the water would have to cover is very long before a leak would appear, the U-shape presents no problem.

The modern deposits bathed by the reservoir are covers superimposed upon the sub-strata, and hence do not offer greater problems either.

Overall treatment of the ground

As an outcome of the surveys carried out, it was thought advisable to carry out preliminary conditioning of the rock area which would later be most directly affected by the tensional pressure bulb transmitted to the ground by the structure load. This procedure, which we chose to call "preconsolidation" was aimed at improving the rock characteristics. It was decided that a preliminary preconsolidation test should be carried out. The lessons thus received enabled us to work out a programme for general preconsolidation :

1. Drilled holes

Ground-plan distribution :

Grid of 3.14 x 3.50 m.

Extent : 20 m downstream from the dam
15 m upstream from the dam

Slope :   15

Slope :   15° to the cystocity

Diameter : not less than $2\,^1/_2$.

With the exception of the holes in the control network, we were not interested in drawing samples, but rather in a drilling process which would ensure a minimum of deviation with a high operational speed.

2. Treatment depth and length of stages :

Treatment depth varied depending on the dam area. It ranged between 50 m in the upper part on the left bank, 30 m along the bed and as much as 40 m in the upper part on the right bank.

Treatment was carried out downward for each layer, following the order established in the operational cycle : drilling, washing and grouting.

We first treated a 7.5 m thick rock roof by means of a low pressure injection (2 kilos/sq cm) of 1:1 proportion grout, a zone which was to be removed when the dam was finally excavated.

The three first layers to be treated, beneath this overlaying bed, were each 5 m thick. In the following ones, down to the depth previously mentioned in each area, the thickness of each layer varied between 5 and 8 m.

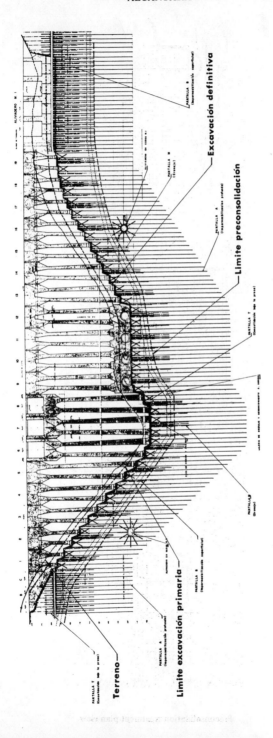

Preconsolidation. Longitudinal profile

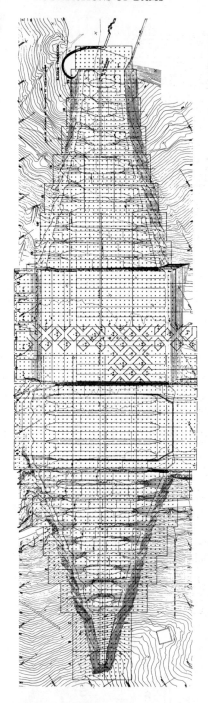

Preconsolidation treatment plan view.

3. Treatment of a group of holes.

The smallest group under consideration consists of 35 holes, consisting of a rectangle of 7 holes in the direction of the dam axis by 5 holes in normal direction, i.e. 7 rows of holes parallel to the maximum dimension and a minimum of 5 in normal direction.

Washing begins with the injection of water and air through the holes in the odd numbered rows parallel to the dam axis and with the opening of the holes in the even numbered rows, thus allowing the water to escape carrying the clay it has picked up while circulating through the fracture. Those where the amount of clay borne by the water is less than 0.5 gr/l may be kept closed. When all the holes in the even numbered rows are pouring out clear water (less than 0.5 gr/l), the air and water injection appliances are fitted to the holes in the even numbered rows and the operation is repeated similar to the description given above, the draining holes this time being those in the odd numbered rows.

Washing is completed with two more successive cycles. Water and air are injected in the even numbered rows perpendiculary to the dam axis in the first cycle, and in the odd numbered rows in the second cycle.

In each stage, injection is carried out through holes or rows of holes parallel to the dam axis, moving upwards, and from upstream towards downstream. No hole less than 20 m from a zone undergoing the washing stage is grouted, whatever the stage to which this hole is subject.

4. Operations to be performed in each hole.

For each group of holes, the operations to be carried out simultaneously in each hole are as follows : it is drilled to 7.5 m depth, the grouting appliance is fitted and the overlaying bed is grouted.

It is re-drilled prior to hardening and the drilled hole is extended by 5 m. This brings it to the definitive treatment layer of the foundation rock. The holes are washed and stoppers are fitted at a level of -7.5 m. Other stoppers with valves are fitted at the mouths of the holes. The first chamber between levels -7.5 m and -12.5 m is then washed out, using pressurised water and air alternatively. The injected water and air, along with the stowage which can be removed by this process, are ejected through the nearby holes which have remained open. When the amount of clay content in the outflowing water is down to less than 0.5 gr./l, the outlet holes are closed and the process is inverted with the holes through which water and air were previously injected now acting as outlets.

Once the required degree of washing had been reached, grout was injected. In the first stage the maximum pressure was 12 kg/sq cm and the concentration of the grout varied between 1:4 and 1:1.

Once the concrete had set, the grouted area was once more redrilled, down another 5 m more, corresponding to the 2nd stage. After a similar washing process, second stage grouting was carried out at a pressure of 15 kg/sq cm. The third grouting stage, carried out after the third drilling and washing stage, was at a pressure of 20 kg/sq cm rising in proportion through the following stages until it reached 30 kg/sq cm.

5. Results

The following figures give an idea of the work achieved :

Total volume of rock treated.............	1,434,500 cub m
Linear metres drilled......	127,750 m
Cement grouted (not counting hole stowage)......	13,430 m.t.
Total time used for rising. ..	71,256 hours
Absorption per meter of hole length	105 kg/cub m
Absorption per cubic metre of treated rock	9.4 kg/cub m
Gain in celerity, converted into elastic modulus	20 % - 35 %

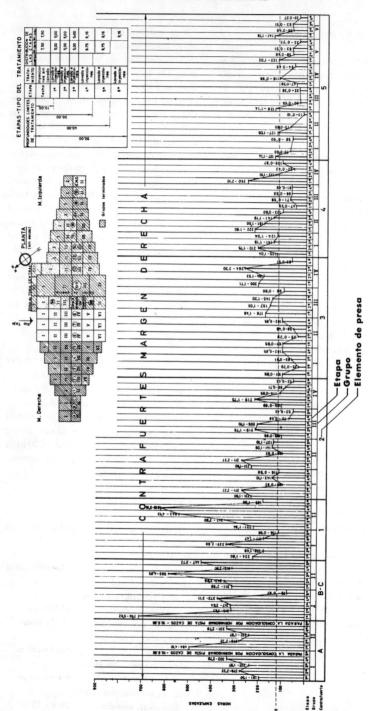

Preconsolidation washing hours graph.

Consolidation groutings of the ground affected by excavation

The aim of the consolidation groutings is to increase the resistance capacity of the supports, by improving their mechanical characteristics and above all, their stability to withstand slipping. This procedure involves reconstructing the compactness of the rock by filling in fractures, cracks and fissures with a grout of sufficient resistance once in service.

It is essential, if this consolidation is to be effective, that the cracks be thoroughly washed out prior to grouting, so as to remove all the clayey filling material which would hamper the passage of the cement grout.

In our case, this aim was satisfactorily achieved, as may be gathered from the description of the treatment given in the preceding section.

Consolidation was therefore limited to trying to seal up any fractures which might have appeared in the rock during excavations. Washing was cut down to the minimum required to clean the hole of dust and detritus prior to grouting.

This consolidation treatment was extended homogeneously throughout the entire area of rock directly supporting the dam, plus a 10 m wide strip upstream and another 15 m wide strip downstream. The thickness of the consolidated zone varied between 7 and 12 m according to the area.

The cement proportions employed varied as a general rule between 4:1 (water : cement) at the beginning of grouting to 1:1 in the holes with greater absorption. In no case was it necessary to inject mortar.

The final figures for this consolidation treatment were :

Treated area . . . .   44,800 sq m

Figure for treated rock
  volume . . . .   513,565 cub m
Holes drilled   7,633
Total drilling length. . . . . . .   98,609 m
Grouting . . . .   1,663,419 kg
Hole filling. . . . .   591,655 kg
Cement absorption per li-
  near m of hole. . . . .   10.86 kg/m
Cement absorption per cub
  m of rock. . . . .   2.08 kg/cub m

Impermeabilization Groutings

Obviously very careful survey and adaptation are required for the watertight diaphragm. This is due to a great many factors of which some depend on the type of structure, but of which the majority depend on the geological nature of the foundation area.

Thorough, reliable reconnaissance work is essential so as to be informed about local geological conditions such as rock quality, permeability, cement absorption, cracks, fissures, etc.

For these reasons, the definition of the diaphragm was worked out on the basis of the results obtained during preconsolidation and consolidation treatment, plus the information regarding permeability provided by the soundings carried out during the actual preparation of the diaphragm itself.

The diaphragm is between 30 and 40 m deep in the central zone and on the right bank, while it reaches as much as 50 m in the upper parts of the left bank.

The diaphragm holes which coincide with the axis of each unit of the dam, were extended to depths of between 80 and 100 metres, thus extending the study of permeability ratings to very deep zones. It was confirmed that there was no need to make the diaphragm any deeper than planned.

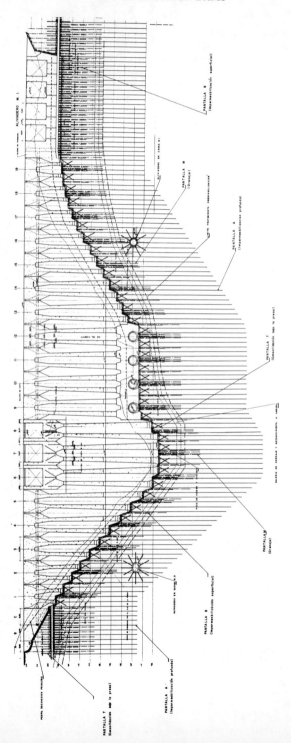

Watertight diaphragm.

The diaphragm consists of a series of holes sloping at 15° upstream; because of the difficulties in drilling as a result of the shape of the dam, they are not in one plane. Both in the ground plan design and also seen vertically, running through the axis of the dam, the pattern is that of a series of fans which describe guidelines which are intended to be so directed as to ensure that the holes of the diaphragm in the deepest areas should not be more than between 3 and 5 m apart, depending on the degree of permeability of each area.

The final figures for the watertight diaphragm are as follows :

Total drillings . . . .   10,699 m
Cement absorbed. . . . .   875 T
Cement absorbed per metre
    of hole drilled . . . .   81.77 kg/m
Cement absorbed per sq m
    of diaphragm. . . . .   24.18 kg/sq m

Water leakages during permeability tests were relatively slight and it was not necessary to inject mortar in any of the holes.

Drainage

The main drainage screen, completed prior to filling the reservoir, consists of 75 mm diameter holes 25 m long, drilled from the upstream end of each unit or joint cavity. In each of these cavities, there are three drains : a vertical one placed in the centre and two side ones sloping at 10 % from the vertical downstream, so that when seen horizontally, they form a line which is clearly parallel to the contact line between the upstream face and the terrain. Following this arrangement, there are 6 drains for each 22 m unit, with an average distance of 4.4 m between them. This screen is extended along the left-bank spillway tunnel, using vertical holes with the same characteristics, 4 m apart.

During loading, additional drainage was introduced in units nos. 3 to 6 and 13 to 15, so as to reduce relatively high sub-pressures (35-40 % of the total pressure) registered at certain points. This drainage consists basically of a strengthening of the main screen with four more drains per unit, and a series of surface holes every 5 m used for draining the bedrock-concrete contact in areas upstream from the dam axis.

A small number of other additional drains were introduced in order to get rid of damp patches which appeared in the bedplates of some of the walls. It can be seen that these leakages derived from earlier survey tunnels lying beneath the dam foundations.

RHEOLOGICAL BEHAVIOUR
OF THE FOUNDATIONS

In accordance with the results of the geological survey, the required geotechnical test programmes were arranged in order to work out the characteristics of behaviour of the formations on which the Alcantara dam is founded.

The tests dealt basically with the following main points :

— Deformability under the effects of the planned structure.

— Resistance capacity and safety coefficient of the dam support.

— Effects liable to be produced by the presence of water from the reservoir.

— Results likely to derive from the envisaged washing and grouting treatment.

The various tests for deformability supplied information based on different, independent procedures, as follows :

— Tests with a plate under the pressures of hydraulic jacks.

— Tests using pressure-meters or 2.1 m diameter flat jacks.

— Tests using elastic wave celerity measurement.

— Tests using a dilatometer in drilling.

On the basis of the results of these tests, it proved possible to draw a fairly accurate picture of the characteristics of the rheological behaviour of the supporting rock foundations, both before and after treatment.

The results of seismic measurements, corresponding to geological observations, enabled us to distinguish between the following zones :

— A surface zone of coverage and weathered rock, pitching slightly. Its celerities range between 1.4 and 1.7 km/s. This was removed during excavations prior to treatment.

— A zone of decompressed rock with more or less open, oxidized and drained cracks with weak silty-clayey fillings deriving from the weathered zone. Average celerities ranged between 3 and 3.8 km/s. The zone proved sufficiently thick in certain parts to call for the study of grouting treatment for the whole.

— A zone of compact rock. This proved to be practically free of defects, of barely variable geological quality and with fairly homogenous geotechnical characteristics. Average celerities ranged between 4.2 and 4.7 km/s. This is a rocky foundation of good quality to provide the basis of a concrete structure.

The following average deformability characteristics were registered for the various rock zones :

— Decompressed rock zone, whose thickness beneath the foundations may range between 2 and 25 m prior to treatment.

According to stratification		Normal	
E	150 T/sq cm	120 T/sq cm	
D	90 "	60 "	
V	0.29 "	0.25 "	

Ditto, following treatment

E	250 T/sq cm	200 T/sq cm	
D	120 "	100 "	
V	0.26 "	0.24 "	

E = Elastic modulus.
D = Deformation coefficient.
V = Poisson Coefficient.

– Deep rock

According to stratification	Normal
E 350 T/sq cm	300 T/sq cm
D 250 "	220 "
V 0.26 "	0.24 "

The minimum intrinsic characteristics registered were as follows :

Cohesion : C = 4 kg/sq cm
Internal friction angle ϕ = 34°

Summing up, then, the supporting terrain is a sort of geological "antique" which, at the surface -where decompression is practically total- has gradually lost part of the good qualities which the rocky core has preserved intact at a certain depth. This was appreciated when we made the two 7.4 m diameter tunnels for temporary deviation, and which were subsequently used as permanent deep drains.

MAIN STRUCTURES

The Dam

Alcantara dam stands at a maximum height of 130 m from its foundations and will create a lake of 3,137 Hm3 at a water level of 218. During exceptional in-flow the level of the reservoir surface may reach 220. The length of the crest at level 223 is 570 metres.

It is an eased gravity type dam, with double buttress. It consists of nineteen elements twenty-two metres wide and two gravity abutments : that on the right bank is 40 m long, with an identical profile, divided into three blocks, (A, B and C). The one on the left bank, which includes the lateral spillway, forms an angle of 20° with the dam axis, downstream. It consists of nine 10 m blocks, an allete wall and a transitional block between the eased gravity and the actual spillway itself.

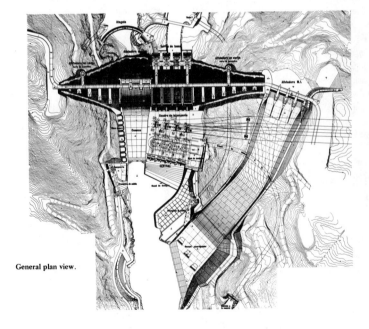

General plan view.

Volume excavated........ 550,000 cub m
Concrete .... 968,500 cub m

The standard element, an isosceles triangle in profile, has the following basic characteristics :

Width .... 22 m
Face batter 0.45
Side face batter................. 0.03
Constant width of winner cavity..... 7 m
Resistant triangle vertex 220.50
Crest level 223
Width of crest .... 7 m

The inner cavity of each element and the space between two consecutive elements, are accessible to inspection via a 1.9 m diameter circular hole which runs throughout the lower part of all the buttresses and also via two tunnels : one at elevation 145 and the other at elevation 209.5 in the lateral blocks, and at elevation 183 in the blocks of the central spillway.

Central blocks numbers 6, 7 and 8, are sealed all along the top in the downstream zone, thus forming the continuous bedplate of the central spillway. The two ends support the spillway encasing walls and each and every one of the piers (semi-piers since they are divided by the inter-block joints) of each waterway.

The foundations of each block ascend in a downstream direction, so as to ensure improved effects of the results of the stresses on the foundations plane. This increases safety against slipping, especially as it means that there is greater discontinuity in possible planes of sub-horizontal diaclasation which might be present underneath the dam. The entire foundations are protected by a ϕ 30 mesh to absorb the retraction stresses and to improve distribution among the weak and strong parts.

Each of the two feet of each buttress is founded on a plane. The off-level between them is made up for in the space of the inner cavity.

The blocks are independent, but in contact throughout with a narrow strip which runs along the entire top where the impermeabilization devices are placed. The devices are :

Upstream zone (from upstream towards downstream)
— Reinforced concrete covering plate with both surfaces which touch the dam blocks lined with two layers of asphaltic sheeting.

— Glass fiber braiding.

— Cup filled with bitumen with electrical resistance heating arrangement.

— 2 mm thick 80 cm wide V-shaped copper strip.

— 20 cm diameter drainage well.

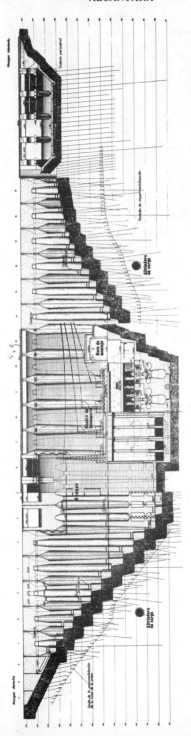

Longitudinal elevation and section through dam axis.

The contact surfaces of any two adjacent blocks have two coats of asphaltic paint, ensuring improved adherence and impermeability.

Downstream zone in the spillway blocks (from downstream towards upstream)

— 22 cm wide P.V.C. foil (SIKA or similar type).

— Copper strip with identical characteristics.

— 20 cm diameter drainage well.

The joint edges in the neighbourhood of the faces are protected with small reinforcements.

For building purposes, each element was divided into blocks which were joined together as shown in the blueprints, calling them construction joints. Subsequently, these joints were grouted, this being done in seasons when the temperature of the dam concrete was at its minimum.

The ledges are two metres thick and sloping :

— The head block ledges slope at 15 % downstream.

— The ledges of the two remaining blocks slope at 10 % upstream.

They are made up of sub-ledges 0.5 m thick, which overlap in the higher parts so as to prevent total continuity in the joints between the ledges.

The cement proportions are 250 kgs. except in the reinforced areas (aprons, beams, etc.) where it is increased to 350 kgs. Six sizes of granitic type aggregate have been used : two of them fine and four coarse, of the following specifications :

Fine sand.	0	—	1	mm
Coarse sand	1	—	5	mm
Pea gravel	5	—	15	mm
Grit	15	—	35	mm
Fine gravel.	35	—	70	mm
Coarse gravel.	70	—	120	mm

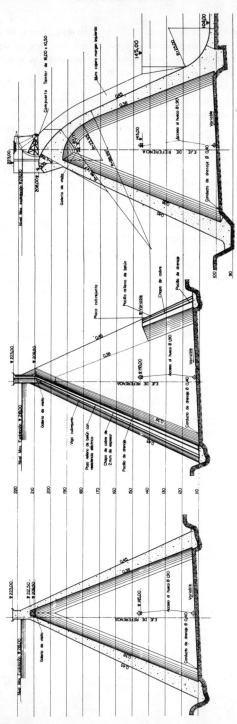

Dam typical unit section

Section through joint between dam units

Spillway unit section

Drainage of the internal spaces of the dam is effected throughout all the blocks by means of 0.8 m diameter conduits placed at the base. Drainage of the upper part of the spillway blocks is effected by means of five 0.4 m diameter vertical holes for each block. This cuts down sub-pressure effects in this area of maximum shear stress.

Spillways

They were designed for a flood peak of 15,000 cub m/sec corresponding to a frequency (T) of 500 years, which is run off after a lopping off of roughly 20 per cent by the reservoir excess height. The remaining 12,160 cub m are removed at 4,160 cub m/sec down the central spillway and at 8,000 cub m/sec down the lateral spillway, for elevation 220 (corresponding to an excess height of 2 m over the normal level at which the reservoir is operated).

For elevation 218, the run-off capacities are 3,100 cub m/sec and 6,650 cub m/sec respectively.

Out-flow has been split off to feed two spillways because of the flexibility of operation that this means and also because of the security this offers should repairs need to be carried out.

Central Spillway

This spillway is situated on the dam itself, and consists of three waterways, each of which are located on one of the elements of the dam.

The waterways are each 16 m wide and the lip is at level 208. They are separated by 6 m wide piers and sealed by gates measuring 16 x 10.95 m.

As already explained, this spillway offers a drainage rate of 4,160 cub m/sec when the reservoir is at level 220 and a rate of 3,100 cub m/sec for level 218.

The power arrangement consists of a hydraulic backwash in a shock-absorption basin with converging channels. The basin has a bedplate lined with 3 m thick concrete slabs, ending at a rising-slope edge or stockpile in order to prevent downstream erosion. This kerbstone ends next to the channels with two lateral deflectors designed for the same purpose.

Drainage of the basin is arranged by means of trenches filled with porous concrete tubes set in gravel. These allow for a run-off downstream from a sill on the bedplate so arranged to provide a low pressure point.

Lateral Spillway

This spillway, situated on the left-bank of the river, consists of the following parts :

a. Control structure with a weir having its lip at elevation 203.5 and with four 15 m wide waterways, separated by piers and fitted with 15 x 15.3 m. Taintor gates.

As already mentioned, its run-off capacity is 8,000 cub m/sec at level 220 and 6,650 cub m/sec at level 218.

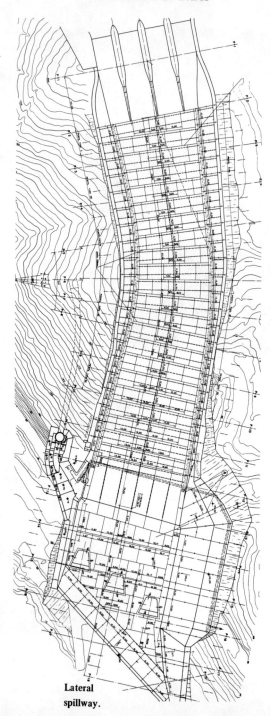

Lateral
spillway.

b. A straight canal with a gentle gradient, albeit sufficient to ensure fast flow.

c. Curved zone with corresponding banking of the bedplate.

d. Fall ramp with a parabolic stretch rising to a tilt of 45° in the lowest part.

e. Shock-absorber basin. The latter includes three blocks for the dispersal of energy, designed to prevent cavitation and anchored to the underlying rock by means of taut cables. The basin ends at a sill whose surface slopes upstream thus preventing erosion at the foot of the structure. It also serves as a cofferdam for the river, enabling the basin to be drained and examined.

A reduced-scale model was used for testing the spillways in two hydraulics laboratories : one in the Hydrographic Studies Centre, Madrid, and the other that of Sogreah in Grenoble.

Dewatering conduits

Two dewatering conduits were designed, one for each bank.

Throughout most of their length, they make use of the two river deviation tunnels and their outlet. The actual drain branch is arranged at the outlet linked up with each tunnel.

Drainage capacity is roughly the same for both. It is 340 cub m/sec for maximum operation level at elevation 218 and 318 cub m/sec for elevation 303.5, this being the level of the lowest weir lip (left bank spillway).

The metallic sealing units have the same dimensions in both cases, although the levels at which they are placed vary slightly. They consist of a 4 x 2.5 m sector gate, controlled at the outlet, and a caterpillar gate measuring 3 x 4 m. The latter are fitted in cylindrical towers placed immediately upstream from the dam. The tower also contains the guides for the metal cofferdam, ensuring water-tightness upstream from the panel which enables the tower to be emptied for servicing the gate and its two guide tracks.

The dewatering conduits have launching springboards which were tested on reduced-scale models in order to ensure the best possible dispersal of energy and a suitable impact zone covering the entire range of reservoir levels.

The right dewatering conduit pours into the centre of the river and the left-hand one makes use of the left bank spillway basin for its run-off, thus ensuring good shock absorption for its energy. In the case of small loads -the last stretch of reservoir dewatering- a reduced-scale model was used to study the shape of the springboards, to ensure that the scanty amount of flow would not give rise to erosion at their foot.

Both bottom drains operated normally during the test launching of the dam.

Structure of the turbine water inlets

The water inlet-points, designed each to cater for a flow of 295 cub m/sec, are situated so that they tally with the axes of elements 9, 10, 11 and 12, located in blocks which are independent of one another and separated from the buttress by a grouted joint.

The feeding structure lip is situated at elevation 155 and the grid structure consists of vertical strips 12 cms apart.

The mouthpiece structure is in the form of a cylinder, truncated by the plane of the grid. The grid panels rest on hydrodynamic profile, horizontal beams, suitably sloping in relation to the current network. An intermediate concrete bulkead is provided in order to reduce the span to the horizontal concrete beams.

Closing and protection are ensured by a caterpillar gate operated by a hydraulic servomotor accommodated in a circular tower. The gate and guides can be inspected through a similar tower which is water-tight and situated upstream from the panel, thus enabling the tower to be emptied.

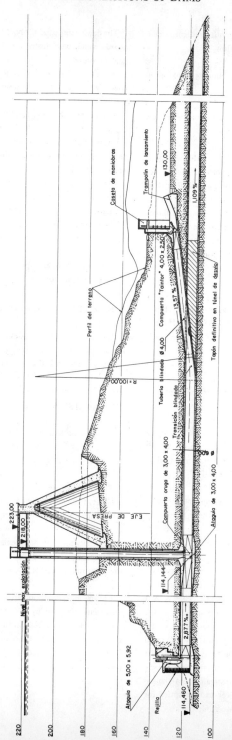

Right bank bottom outlet.

Ancillory defensive works in the reservoir or downstream

Downstream from the central spillway basin protection has been provided for the right-hand bank. Coating has been provided for the rock in the area near the outlet of the right-hand dewatering conduit, which is affected by the waves deriving from the central spillway and the out-flow of the dewatering conduit itself with scanty load.

Downstream from the dewatering conduit, the hillsides have been gunited up to the level of the maximum foreseeable flood. This affects the entire area facing the left-bank spillway shock-absorption basin, which is thus protected against the waves produced by the latter during out-flow when said waves produce a frontal attack. Two bulk concrete walls have been built horizontally at two different levels, half way up the hillside, forming part of the protection system against the above-mentioned effects. A section of rock between sub-vertical faults transversal to the ri-ver, which underwent a landslide, has been protected and held in position by a concrete wall anchored with stressed cables.

METAL PARTS

Central spillway gates

There are three 16 x 10.95 m sector gates operated by chain and winch. The gate edge is arranged to rest at elevation 207.35.

Upstream there is a metal cofferdam divided into 6 elements and operated by a gantry crane.

The gate hinge axes are supported by metal beams held to the piers by means of prestressed cables.

These cables transmit the gate stresses to the upstream part of the pier.

Lateral spillway gates

There are four 15 x 15.42 chain and winch-operated sector gates. The gate edge rests at a level of 202.88.

Upstream there is a metal cofferdam divided into 8 elements and operated by the gantry crane.

The gate hinge axes are supported by metal beams held to the piers by means of prestressed cables in a similar way to that described for those of the central spillway.

Dewatering conduit gates

The two -right and left hand bank- dewatering conduits have similar gate arrangements. The dimensions are the same for both although the levels at which they are placed vary slightly.

They consist of a regulation sector gate 4 m wide and 2.5 high, situated downstream, and with an approximate load of 98 m.

The protective closing system, upstream, consists of a caterpillar gate 3 m wide and 4 m high with an approximate load of 103 m. This was designed to open and close under load where necessary, although it is normally used under balanced loads.

The 3 x 4 m cofferdam may be situated upstream from the gate. It is watertight upstream from the panel, thus allowing for the cylindrical tower to be emptied and the gate and its guides to be inspected.

It was not thought necessary to have a grid upstream from the dewatering conduits because it seemed there was very little danger of obs-truction.

Water intake point gates

Closure and protection are ensured by means

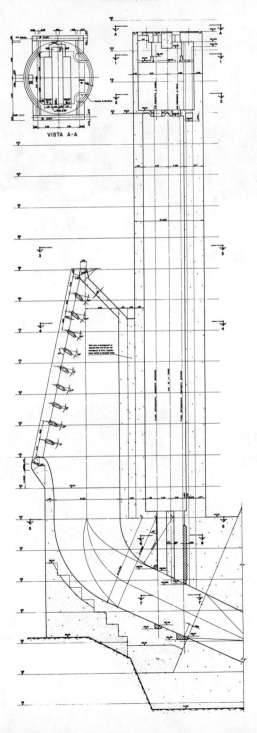

VISTA A-A

Dam and Power Station.
Conduit section.

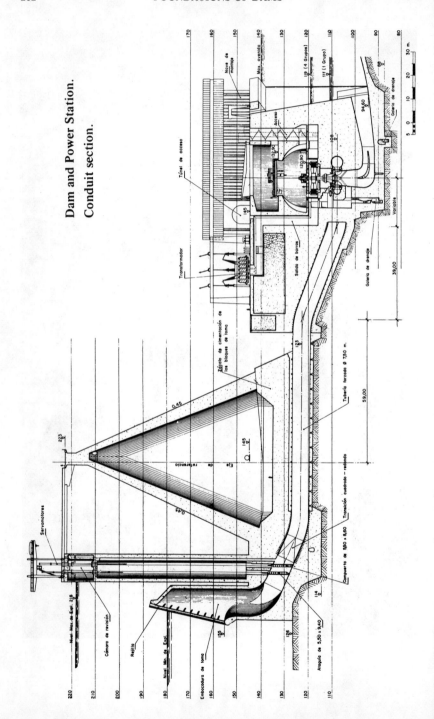

of a caterpillar gate 5.5 m wide and 8.6 m high, operated by an oil piston with a servomotor. The level at which it rests on the sill is 125.817. The gate was designed for a reservoir level of 220, i.e. for 94.2 m of water load.

The gate is arranged so that it can close both for the machine-racing flow and for the free jet flow - in an extreme case where there might be a break in the pipes.

Upstream from this service gate, there is an emergency gate 5.5 m wide and 8.75 m high, watertight upstream from the panel, thus allowing the tower to be emptied and the panel and service gate guides to be inspected. In cases of emergency closure and a possible breakdown of the main gate, the emergency or security gate is designed to close under the same conditions as the former.

Turbine in-take pipes

Plating begins at the point of transition between circular and rectangular cross-sections, coinciding with the joint between the dam buttress and the in-take structure.

This transition section and the vertical elbow are the only parts encased in the surrounding concrete. The remainder of the pipe (diameter 7.5 m), is free as far as the point where it enters the power station, being housed in a visitable tunnel running through the cavity existing in the dam element axis. This guarantees both the piping and the dam against possible, dangerous effects of intersticial pressure likely to be transmitted mutually.

In the free stretch, every 3 m the piping is supported by rests on rollers to facilitate longitudinal movements deriving from dilation. Since the rock underlying the piping has undergone deformation due to the weight of the dam, which is not the case of the piping encased in the power station, a joint has been introduced immediately preceding the latter stretch, to provide for transversal movements as well as the longitudinal movements due to temperature changes.

CONSTRUCTION

Criteria for the deviation

The measurements and gagings carried out in the river over the previous 15 years (1947-1962), led us to foresee annual floods of between 1,000 and 1,500 cub m/sec which might appear during the half year between October and March. Every two years these floods were likely to reach peaks of over 3,000 cub m/sec.

This hydraulic pattern meant rejecting any idea of totally diverting the river, an arrangement which would have enabled us to keep the working area dry throughout the entire period when the dam and power station were being built. However, this would have meant having exceptionally large and uneconomical cofferdams and the corresponding deviation tunnel.

Furthermore, such was the narrowness and depth of the river in the stretch in question, that there was no possibility of arranging for partial deviations of the channel. Hence it was decided to opt for complete deviation during the low water season. This would ensure that the working area could remain dry for six or eight months of the year, while being foreseeably flooded during the rest of the time.

During flood periods, the plan was that the water should pass through the dam. Operations were coordinated in such a way that work was transferred at the end of the dry season from the lower to the upper zones. This prevented altering the concreting programme, and avaided hindrance to the general progress of the work. The flood periods, during which the water passed through gaps left in the dam, were used for closing off the deviation tunnels and preparing them to serve as dewatering conduits.

Deviation works

The main deviation works were as follows :

1. Tunnels. Two tunnels 550 m long and 6 m diameter were drilled after coating with concrete. These were subsequently converted to use as permanent dewatering conduits, as explained above.

2. Cofferdams. The upstream cofferdam provided a maximum deviation of 500 cub m/sec and stood 37 m high. This was a thin concrete arch dam, especially designed so that any rise in the river would flow over the crest. This crest is 140 m long.

3. Passage through the dam. Since it was likely that considerable floods might occur during the building of the dam and that these could not be easily deviated through the tunnels, three large passageways, measuring 6.6 x 12 m were provided, in three of the dam elements on the river bed. The central river bed element remained at foundation level, providing improved drainage throughout its 22 m width, pending completion of the rest of the dam to a suitable height where, if necessary, the three main drains of the project could be used successfully.

Incidences during construction

The high irregularity of the river in the Alcantara area follows from study of the available hydrological data. In March, 1947, a flood was registered with a peak of almost 10,000 cub m/sec. In subsequent years, in 1949 and 1950, years of considerable drought, the river flow in August and September was only of 5 cub m/sec.

Although, at the beginning of Alcantara, considerable control of the river had already been achieved, thanks to the reservoirs built between the Tagus headwater up to Valdecañas, there were scant controls on the rivers Tietar and Alagon, which carry the run-off from the Gredos and Gata mountains, and contribute about 50 % of the Tagus waters at Alcantara.

In fact, the preliminary rock-fill cofferdam holding 150,000 cub m, required for deviating summer flow and thus enabling the final concrete cofferdam to have its foundations laid and be built, was partly destroyed on several occasions; it was not until 1967 that the main concrete cofferdam could be completed.

In 1968, there were no rises in the river level worthy of mention. At the beginning of the year, the river had a flow of about 1,100 cub m/sec which poured over the foundations of the central unit of the dam on the bed and through the 6.6 x 12 m holes in the two adjoining units. Immediately the flood season was over, work recommenced rapidly on construction of the unit in the centre of the river bed.

In 1969, there was a prolonged period of high water at the beginning of the spring. Flow surpassed 4,000 cub m/sec but it poured easily through the three 6.6 x 12 m holes.

OPERATION

Surveillance and Monitoring Arrangements

1. The Alcantara dam is the highest to have been built of its type so far. The foundations conditions, mainly as a result of the preconsolidation treatment provided, may be described as excellent. There are only local problems of sub-pressure through the system of cracks, and even these problems are alleviated by the large draining surfaces provided at the base of the dam.

The central installations have been arranged as follows :

— Safety or emergency devices cover each and every one of the units in such a way that they operate independently. A distinction exists, however, in the number and variety of devices for each unit or element, mainly in terms of their height, with maximum control being provided for elements 5 to 13, medium for 3, 4, 14, 15 and 16 and minimum in the remainder.

— Research centres on a detailed examination of elements 5 and 7, these being these of greatest height among those of the normal type and of spillway type respectively. This also applies to the bedplate containing the piping driven into element II.

2. External action is measured with the following apparata devices :

— 1 limnigraph for recording reservoir water level.

— 60 fixed piezometers for measuring sub-pressure in the foundations of all the odd numbered elements. These are generally placed where the concrete contacts the rock, but some have been placed at a depth of 15 m.

− 1 weather-recording unit complete with thermometer, barometer, anemometer and rain gauge, for registering environment conditions.

− 6 thermometers for measuring water temperature at different levels of the reservoir.

− 20 thermometers for measuring at temperature inside the cavities in the elements.

3. Safety control of loading effects includes :

a. Measurement of horizontal displacement :

− A geodesic system, with 10 vertices and 40 facing signals.

− A collimation system, with 2 vertices and 22 facing signals.

− 3 inverted plum-bobs, anchored at a depth of 15 m beneath the foundations of units 6, 7 and 8, with three measuring points each.

− 6 ordinary plum-bobs, in units 5, 9, 10, 11, 12 and 13, with two measuring points each.

− A system of gages for measuring the joints on the crest of the dam, with a triangular based measuring system for all the interunit joints.

b. Measurement of vertical displacements :

− A levelling system, providing control of 27 points along the dam crest.

c. Measurement of rotation :

− 44 clinometer bases situated in the foundations of almost all the units, and at various heights in units 5 and 7.

d. Measurement of compression :

− 55 string tensionmeters placed in the foundations of units 5, 7 and 11.

e. Seepage gaging

− This is carried out in the drains, joints and final collection points.

4. The following apparata are used, in addition to some of the foregoing, for research purposes :

a. In unit 5 :
− 42 string thermometers for measuring the temperature of the concrete.

− 56 string extensimeters for measuring warping, distributed among corrective devices and vertical rosettes.

b. In unit 7 :
− 82 thermometers.
− 112 extensimeters.

c. In unit 11 :
− 10 thermometers.
− 13 extensimeters.

5. Data collection and visual inspection are the task of a team of five watchmen who tour the inspection tunnels daily and who are responsible for maintenance tasks.

The normal surveillance programme includes the following frequencies for apparatus readings :

− Continuous reading :
level of reservoir; meteorological data.

− Twice a week :
plum-bobs.

− Weekly :
clinometers, piezometers, tensionmeters, water thermometers and air thermometers.

− Fortnightly :
extensimeters, concrete thermometers, seepage gages.

− Monthly :
collimation, joint movement.

− Half-yearly :
geodesic readings, levelling.

Behaviour during loading and operation -

1. The first filling of the reservoir began at the end of August, 1968. Loading occured fairly rapidly since the date coincided with a period of exceptional in-flow. By January 1970, the level of 216.6 had been reached (1.4 m below the maximum normal level). Subsequently, the reservoir remained at roughly the same level until June 1970. An idea of the good behaviour of the dam under these stresses can be gathered from the following data registered for the situation of maximum filling :

— Sub-pressures in the drainage screen remained as a rule at less than 15 % of the total pressure.

This figure was surpassed at only 4 of the 18 points registered, although it did not reach 40 % at any of them.

— Sub-pressures downstream from the drainage screen generally remained at less than 10 per cent of the total pressure. Only 5 of the 23 points registered surpassed this figure, while none reached 22 %.

— Seepage in joints between units reached a maximum figure of 33 l/min.

— Horizontal displacements in the direction of the river reached a maximum reading of 19 mm on the crest of unit 7.

2. The period between June 1970 and the present time, may be taken as one of normal operation. The reservoir level has varied between 181 and 211. The following data, read in March 1972, when the reservoir was at level 211, give a fair picture of the dam's behaviour.

— Sub-pressures in the drainage screen remained generally at less than 15 % of the total pressure. Only 5 of 19 points checked rose above this figure, and yet remained below 28 %.

— Sub-pressures downstream from the drainage screen remained generally below 10 % of the total pressure. Only 2 of the 24 points checked rose above this figure, and yet remained below 22 %.

— Seepage through the drains reached a maximum of 173 l/min.
— Seepage through joints between units reached a maximum of 212 l/min.

— Horizontal displacements in the direction of the river registered a maximum of 18 mm on the crest of unit 7.

POWER STATION

The power station structure at the foot of the dam was designed as a mass, so that the considerable excavations which were required for it, with banking of the cut in the rocky core on the left bank to a height of 85 metres, should be secured by the upstream and downstream structures of the axis of groups operating as core buttresses. Furthermore, compressions and a mass were introduced at the foot of the dam, thus encouraging an increase in the capacity of the foundations to resist passive thrusts.

The induction pipes, 34.75 m long, immediately below the cone are provided with a tongue and central panel which cut down turbulence at the outlet. The outlets of the induction pipes have been designed with the required cofferdams to ensure that they can be kept dry for inspection. These cofferdams are operated by means of a gantry crane which runs along the platform at level 145.

There is a road leading to the power station along the left bank and access is through a tunnel running beneath the left bank spillway.

The alternator platform may be reached through four inset wells and by easing the closure on the downstream side of the power station, where the corresponding stairs and lifts are situated.

The most important significant levels in the power station are as follows :

92.2 m a.s.l. the lowest induction level.

108   ”   average level for distribution.

112   ”   control unit, servomotors, refrigeration water intake, etc.

116   ”   stator bedplates, bottom of alternator well, etc.

120.9   ”   exciting-dynamo plant.

The 15 KV bar outlet runs through wells situated in the concrete core on the upstream side of the power station as far as the transformer station at level 145, between the power station and the dam.

These transformers raise the voltage from 15 to 400 KV, this being the adopted transmission voltage.

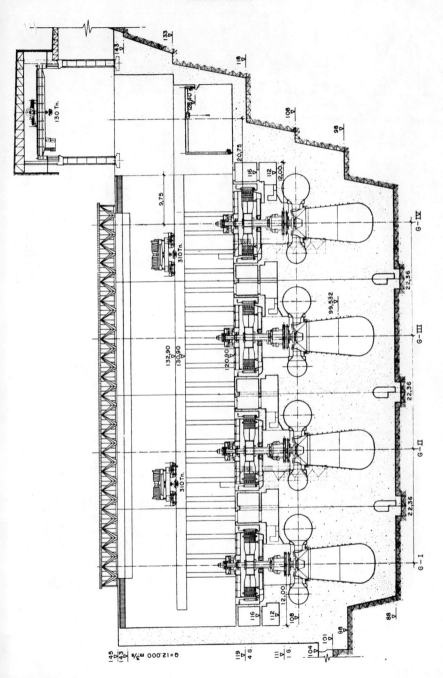

Two pairs of gantry cranes were installed for the assembly and maintenance of the heavy machinery. One consists of twin 130.15 ton, 18 m span cranes running through the assembly workshop set up at level 145, thus providing access for materials to level 20.9 (in which an area was also allocated for the assembly of the rotor) on a truck which approaches the gantry crane area orthoganally overlapping with another couple of twin 310/15 ton, 15 m span cranes which run along concrete corbels specially arranged for this purpose at level 132.9.

The machinery has the following characteristics :

Four Turbines

Maximum gross head. . . .	108 m
Minimum gross head	58 m
Nominal gross head	81 m
Maximum flow per set. . .	293 m3/sec
Nominal flow per set. . . .	289 m3/sec
Nominal power	275,000 H.P.
Maximum power	316,000 H.P.

Nominal speed.	115.4 r.p.m.
Racing speed	245 r.p.m.

Four alternating-current generators

Nominal power	286.8 with cos = 0.8 and heating 80° C
Voltage	15 + 10 KV
PD2.	47,500 Txm2
Short-circuit ratio	1.1

Four Transformers

Type	TPA F 222,000/400 KV
Nominal under power continuous operation. . . .	255,000 KVA
Transformation ratio with no load·.	15/400 ± 2.5 ± 5 % KV
Connection	Triangle/star with accessible neutral
Connection group	Yd 11
Heating	in copper 55° C in oil 50° C

MATTMARK - A DAM FOUNDED ON

ALLUVIAL AND MORAINIC DEPOSITS

By Dr. Bernhard Gilg[1]

1. General situation

The Mattmark hydroelectric scheme is located on the Saas valley, in
Valais, Switzerland. The Reservoir of Mattmark, which provides the
seasonal storage capabilities of the scheme is formed by impounding
the Saaser Vispe river, a southern affluent of the Rhone river. The
total drainage area of the scheme is 90 km2, 60% of which corresponds
to diversion from lateral valleys. The usable storage is 100 million
m3.

In spite of the relatively small storage capacity, a total annual energy production of 600 million Kwh can be achieved in virtue of the available head of nearly 1500 m. This high energy production requires the
best possible use of the water available, practically no water losses
being allowed to occur. For reasons of economy it was, therefore, desired to reduce the amount of seepage from the Reservoir as much as pos-

1) Head of Power and Irrigation Department of Electro-Watt, Zurich

sible, beyond the extent needed for reasons of the dam stability.

Fig. 1 shows a foto-assembly of the dam and of the lateral moraines, characteristic for this site, which divide the valley into platforms.

Photo assembly of the Mattmark dam

Fig. 1

2. Description of the subsurface formation

2.1 Rock

The underlying bed is composed of cristaline rocks pertaining to the Monte Rosa nap. The predominant rock types are granite, gneiss and prasinite. At the dam site the rock layers strike in direction N 60° E and dip 40° towards NW. From this local morphology it follows that, dominantly, the layer heads appear in the left abutment and the layer faces in the right abutment zone. For this reason it also follows that joints are better closed in the left abutment zone and that weathering effects, resulting from water circulation, are stronger on the opposite abutment. However, since the valley flanks were subject to intense mechanical action of the glaciers during the last glaciation period, the weathered zone is nowhere thicker than approx. 10 m and is, therefore, of minor importance.

Fig. 2 shows a longitudinal and a cross section of the local geological features of the dam site.

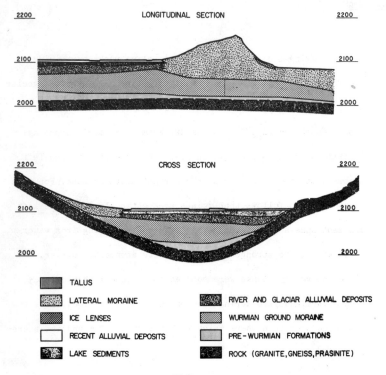

FIG. 2

Geologic section

2.2 Deposits

Overlying the sound rock surface, the valley trough is filled with loose material with a thickness of up to 100 m. As shown in Fig. 2 the deposits are, from bottom to top:

- deposits from the interglacial period
- ground moraine from the latest glacial period consisting of fine to coarse grain size material, layered in a very heterogeneous way
- alluvial gravel deposits of the post-glacial period, which were never subject to the glacier weight and therefore are not as dense as the ground moraine.
- at the surface, a loose silt lake deposit, formed during the period of time when the Mattmark plane was the bottom of a lake impounded by a lateral glacier.

An adequate characteristic of these materials is supplied by the grain-size distribution curves in Fig. 3

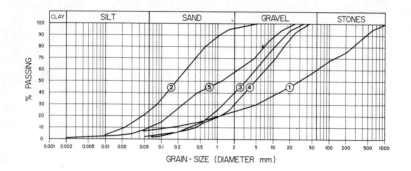

REPRESENTATIVE GRAIN-SIZE DISTRIBUTION OF FOUNDATION MATERIALS

I LATERAL MORAINE — 2 LAKE SEDIMENTS — 3 RIVER AND GLACIER ALLUVIAL DEPOSITS
4 WURMIAN GROUND MORAINE — 5 PRE-WURMIAN FORMATIONS

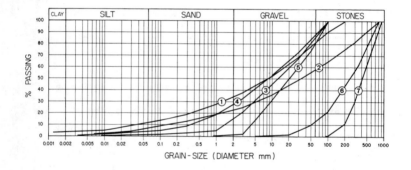

REPRESENTATIVE GRAIN-SIZE DISTRIBUTION OF FILL MATERIALS

I CORE — 2 SHELL — 3 DRAINAGE ZONE UPSTREAM — 4 FILTER
5 DRAINAGE ZONE DOWNSTREAM — 6-7 RIP-RAP

FIG. 3

Grain size distributions

The different subsurface materials may be characterized by the follo-

wing parameters:

Material	Permeability*	Compression modulus	USCS clas- sification	Internal friction	Cohesion
	k (cm/s)	ME (kg/cm2)		tg φ	c (kg/cm2)
Pre-Wurmian formations	10^{-2} to 10^{-3}		SW		
Wurmian Ground moraine	10^{-1} to 10^{-3}	750	GW	~ 0.80	-
Alluvial Deposits	10^{-2} to 10^{-4}	300	GM	0.68	0.3
Lake sediments	10^{-3} to 10^{-4}		SM		
Lateral moraine	10^{-2} to 10^{-3}		GW	0.90	-

*(see Fig.no. 4)

The permeability was determined by means of a large number of in situ

tests as shown in Fig. 4. At each test location drawdown and pumping

tests were carried out. It is a well known fact that the pumping

tests provide, in general,higher values than the drawdown tests,

the difference in results increasing with the amount of fine mate-

rial. The likeliest values of k were determined using the formula:

$$k = \sqrt{k_d \cdot k_p}$$

where:

k_d: permeability from drawdown test

k_p: permeability from pumping test

The results obtained in this way vary over a considerable range due

to the heterogeneity of the deposits.

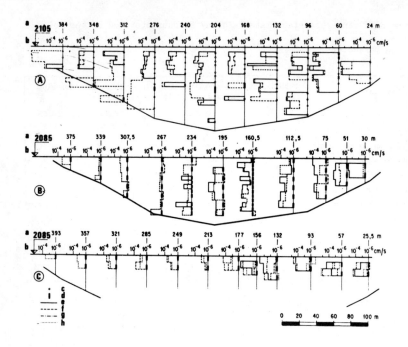

Fig. 4

Permeability test results

(A) Permeability in the middle of the curtain before grouting.
(B) Permeability in the middle of the curtain after grouting.
(C) Permeability at sections upstream and downstream following grouting.
(a) Spacing of borings measured from the end of the grout curtain on the right bank.
(b) Coefficient of permeability k.
(c) Permeability test at a point within a bore-hole.
(d) Permeability test on a several meters long bore-hole reach.
(e) Pumping tests either downstream or in the middle of the curtain.
(f) Drawdown test downstream or in the middle of the curtain.
(g) Pumping tests upstream of the curtain.
(h) Drawdown tests upstream of the curtain.

The materials used in the dam construction were obtained from
these different deposits. Thus:

core material	obtained from the lateral moraines with limitation of the max. grain size
filter material	obtained from the alluvial deposits
drainage layer material	obtained from the alluvial deposits by elimination of fine grain size material through washing
shell material	obtained from the lateral moraines
rip-rap material	results from the excess (large grain size) during production of the core material by use of a Wobbler-Installation

3. Choice of the dam layout and type

3.1 Construction materials

Due to the thickness of the deposits only an earth or rockfill
dam could be considered avoiding the need for a deep excavation.

Without such an excavation the treatment of the pervious foundations was a major problem which had to be solved. Also, since only
moraine material with very little clay fraction was available, a
wide core was required.

Finally, for reasons of economy, the lateral moraine which cros-
ses the valley, was to be incorporated into the dam body but could
not be used to replace the core zone in virtue of it's excessive
perviousness. Therefore an inclined core proved to be the most
adequate solution.

Fig. 5 represents a typical cross section of the dam and shows
the different zones. An upstream shell is not necessary, since
the core material has a friction angle as high as 42° and a relatively low pore pressure coefficient (smaller than 5%).

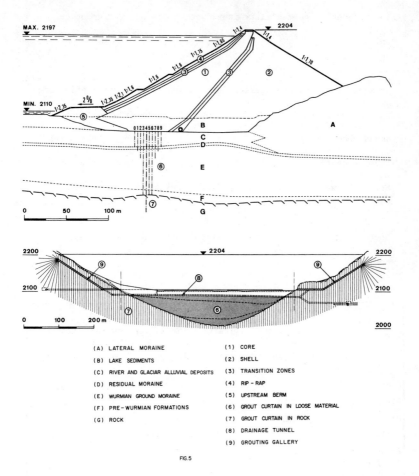

MAX. 2197

MIN. 2110

2204

FIG.5

(A) LATERAL MORAINE
(B) LAKE SEDIMENTS
(C) RIVER AND GLACIAR ALLUVIAL DEPOSITS
(D) RESIDUAL MORAINE
(E) WURMIAN GROUND MORAINE
(F) PRE-WURMIAN FORMATIONS
(G) ROCK

(1) CORE
(2) SHELL
(3) TRANSITION ZONES
(4) RIP - RAP
(5) UPSTREAM BERM
(6) GROUT CURTAIN IN LOOSE MATERIAL
(7) GROUT CURTAIN IN ROCK
(8) DRAINAGE TUNNEL
(9) GROUTING GALLERY

Longitudinal and cross-sections of the dam and foundations

The foundation permeability is reduced by means of a grout curtain.

An accessible drainage gallery designed to collect the entire seepage accross the dam and foundations is located at the toe of the
downstream drainage zone. Due to it's action, the highest water
level elevation in the downstream shell raises only a few meters
above the water level in the gallery, and no uplift occurs.

The average material characteristics of the different dam zones
are:

	unit weight of dry soil	Porosity	Permeability	Internal friction	Cohesion
	γ_d (t/m3)	n (%)	k (cm/s)	tg φ	c (kg/cm2)
Core	2.48	17	2×10^{-5}	0.90	0.20
Filter	2.20	22	3×10^{-4}	0.83	0.80
Drainage	2.04	27	10^{-1}	0.80	0.75
Shell	2.45	18	very pervious	0.92	0.00
Alluvial deposits (as far as rele- vant for the dam stability)	2.14	24	10^{-3}	0.68	0.40

Average material properties

3.2 Stability conditions

The stability analysis of a dam such as Mattmark has necessarily

to take into account the foundation conditions. Contrarily to con-

ditions prevailing on the downstream side where in all load cases

the slope is practically the only determinant factor of the dam

stability, the critical slide surfaces on the upstream side dip

into the alluvial foundation deposits down to a depth of about

20 m. Therefore it was necessary to provide a toe embankment. The

critical load conditions to be considered are:

- normal case: rapid drawdown of storage level
- exceptional case: rapid drawdown associated with earth-
                     quake

The minimum computed values of the safety factor F were:

- load condition		safety factor F	
		upstream slope	Downstream slope
During construction (until 1967)		1.50	1.45
Full storage	without earthquake	1.60	1.52
	with earthquake	1.35	1.29
Rapid drawdown	without earthquake	1.33	1.52
	with earthquake	1.03	1.29

Earthquake effects were replaced by an equivalent statical load

corresponding to an acceleration of 10% g. Although a safety fac-

tor of 1.03 is relatively low it may be considered as acceptable
since the corresponding event is very unlikely to occur and also
since no catastrophic conditions may occur when the Reservoir has
been depleted. The actual safety factor is also likely to be higher
than the value given above since conservative values of the material parameters were used in the analysis.

4. Foundations treatment
_____

It is possible that no treatment of the foundations would have been
necessary for stability reasons. However, without such a treatment,
the total seepage corresponding to full storage would be of the order
of 1 to 1.5 m3/s, i.e. the lake would loose 20 to 30 million m3 annual-
ly. This value is approximately 25% of the storage capacity and corresponds to 80 million Kwh peak energy or a 2 million US $ annual loss
approximately. These figures show the need for a foundation treatment.

A cut-off wall or a pile wall appeared to be economically unfavorable
and technically problematic in virtue of the great depth, considerable
overburden due to the dam construction and of the large size material
occuring in the foundations. Thus, decision was in favor of a grout
curtain. This curtain extends over the entire alluvial and moraine zones between the core zone base and the subjacent rock and even partly

penetrates into the rock, mostly in the abutment zones. The effective
depth of the curtain is of about 100 m over the entire dam length, as
shown in Fig. 5.

In the rock zone the grout curtain consists of a single row of injection holes below the injection gallery and of 2 to 4 parallel rows,
fanshaped, between the gallery and the dam core. In the alluvium material, the grout curtain consists of a number of parallel rows decreasing from 10 at the top of the curtain, to 4 at it's bottom.

Fig. 6 shows the detailed disposition of the grout curtain. The spacing between injection holes in each row is 3.0 m and the distance
between rows is 3.5 m. This means that each injection hole controls a
horizontal area of 10.5 m2.

Different gels were used as grout material. After preliminary tests
the following 2 compositions were selected:

	Clay - Cement Grout		Bentonite - Silicate Grout
Water	850 1	Water	950 1
Clay	300 kg	Bentonite	100 kg
Cement	80 kg	Silicate	16 1
Silicate	12 1	Phosphate	14 kg
	Composition of the unit volume (1 m3) of Grout		

The distribution in the boreholes is shown in the following Table:

Elevations	Number of the injection row				
	0/9	1/8	2/7	3/6	4/5
Phase I : Clay-Cement Grout	m3 of grout per m of drillhole				
2090 - 2070	3	1,5	3	-	-
2070 - 2065	-	3	3	-	-
2065 - 2055	-	-	3	1,5	-
below 2055	-	-	-	3	1,5
Phase II: Bentonite - Silicate Grout					
2090 - 2070	1,5	1,5	1,5	3	3
2070 - 2065	-	1,5	1,5	3	3
2065 - 2055	-	-	1,5	1,5	3
below 2055	-	-	-	1,5	3

Distance between injected drill holes along
sealing surface (m)

The average quantity of injected material was 3.7 m3 of gel per m of
borehole length, i.e., approx. 0.35 m3 of gel per m3 of treated soil.
Since the natural porosity of the soil was 30% prior to the injections,
heave occurred resulting in a maximum upward vertical displacement of
60 cm as shown in Fig. 6. This heaving, clearly indicative of super-

saturation, proved advantageous since, during loading due to the

dam construction, part of the grout material was compressed into fi-

ner pores of the foundations which resulted in an increase in imper-

viousness during the last years of the construction.

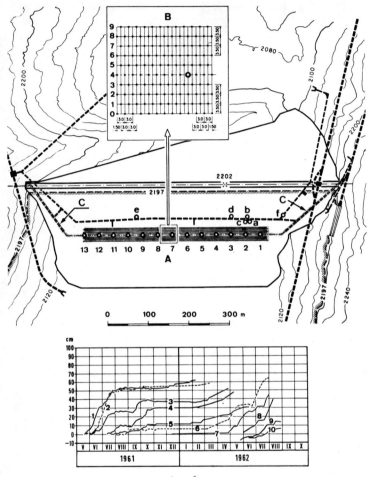

Fig. 6

Position of the grout curtain and ground heaving during grouting

 (A) Impervious section

 (B) Table of borings with number of rows

 (C) Grout curtain within the rock

(a-f) Sampling positions in the drainage tunnel for the chemical
    analysis of infiltration water

(1-13) Measurement points during the heaving of the ground following
    grouting

The global characteristics of the grout curtain are summarized in

the following Table:

Sealed area in loose material zone	21500 m2
Loose material volume treated	516000 m3
Sealed area in rock zone	65000 m2
Total drill hole length in loose material zone	72000 m
Injected length of drill holes in loose material zone	49000 m
Total drill hole length in rock zone	12000 m
Volume of injected gel in loose material zone	215000 m3
Materials Clay	36000 t
Cement	15000 t
Bentonite	11000 t
Silicate and Phosphate	5000 t
Grouting pressure range in loose material zone	20 to 25 atmospheres
Weight of cement injected in rock zone	7000 t
Average take in rock zone	580 kg/m
Grouting pressure range in rock zone	10 - 40 atmospheres

Characteristics of the Grout curtain

These data show that the grout absorption in the rock zone is exceptio-

nally high for a sound gneiss. It is presumed that at a greater depth

and mostly in the right abutment zone extensive fractures occurred

which were closed by grouting.

As a complement to the main injections, secondary injections using silicate were carried out in the upper alluvial zones. These secondary injections were not very efficient and did not contribute much to the curtain imperviousness.

5. Dam and foundations behaviour during construction

5.1. Before occurrence of cracks (1960-1965)

The most interesting results of measurements at the Mattmark dam correspond to the determination of settlements and of horizontal displacements. Pore pressures did practically not develop in the dam core and no evidence of exceptional uplift effects could be observed in the dam foundations.

The settlements measured, until the end of 1965, indicate a total compression of the natural subsurface formation of at most 1.75 m below the drainage gallery and of 1.65 m in the central section of the dam below the crest, as shown in Figures 7 and 8.

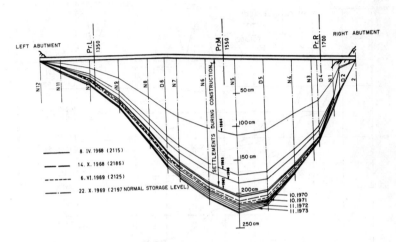

DRAINAGE GALLERY SETTLEMENTS

FIG. 7

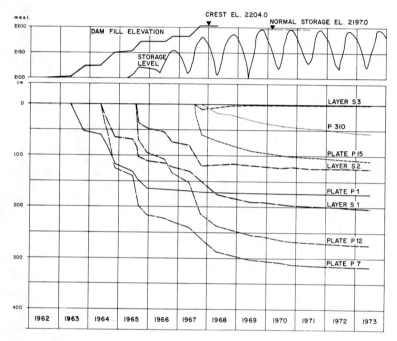

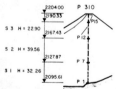

SETTLEMENT RECORDS: VERTICAL GAGES
LOCATED IN SECTION 1550 AT DAM AXIS

FIG. 8

At this time the embankment weight was 85 m and the total cumu-
lative compression of the central dam shell was approx. 3 m. This
relatively high value results from the method of construction of
the shell in 1 m thickness layers and compaction by means of the
hauling equipment only.

Settlements occurring in the dam core zone, compacted by 6 passes
of a 80 ton weight pneumatic roller, were much smaller. However,
since the inclined core rests upon the dam shell it had to follow the motion of the latter in a rotational movement around it's
toe. The extent of the core zone motion is increased by the differential settlement of the foundation itself, these settlements
being larger under the shell zone than under the core.

The motion of the core was the reason for the occurrence of 1
major and 2 minor cracks in the right abutment core zone, which
became apparent at the end of the year of 1965 shortly after the
interruption of the construction work at the beginning of the
winter recess.

5.2. After occurrence of cracks (1966-1967)

The main crack can be clearly visualized on Fig. 9. It is located in a vertical plane forming an angle of about 45° with the

dam axis and approaches the abutment towards downstream. The total depth of the crack, determined by means of a vertical shaft
driven into the core zone (see Fig.10), was 7 m. The distance
between the crack and the dam - abutment contact was 30 m on the
upstream and 10 m on the downstream face of the core.

Fig. 9   Fig. 10

Dam surface showing a crack

The vertical plane containing the crack intersects the foundations in the sound rock zone and not in the zone of deposits,
i.e., on solid base.

All these features indicate that the crack is the result of flexural tensile stresses induced by the movement of the core described earlier. The flexural stresses are particularly important
in the right abutment zone since, there, a strong variation of
settlements occurs over a relatively short distance along the
dam axis.

Conditions in the left abutment zone are quite different since
the lateral moraine provides a smoother transition between the
alluvial and the rock foundations. The total crack width may
have been of about 30 mm at the surface.

The embankment construction was completed during the years 1966
and 1967 proceeding from the left bank. On the right abutment,
construction in the zone where the cracks had occurred was de-
layed as long as possible. The major part of the settlements occurring during these 2 years of the construction could, therefore, take place with participation of the old cracks. No additional cracks were formed during this period although the settlements attained values as high as 30 cm. During the last construction phase the zone of cracks was excavated down to a depth

of 7 m and, finally, the breach was filled.

Extensometers installed at 3 different locations of the right
abutment zone still indicated a displacement of at most 50 mm
over a length of 6 m at a place where the major crack had previously occurred. However, this displacements could be bridged
by the material plastic behaviour and no evidence of crack formation could be observed.

Even in the event of an undetected internal crack formation, the
resulting condition would not be very serious, since the moraine
material, in virtue of it's grain size distribution, possesses
self healing properties. Also, the extensive drainage and filter
zones would completely avoid the wash-out of core material.

The horizontal displacements which took place until the end of
the construction period were already important and reached values up to 32 cm at the upstream face (towards upstream) and up
to 40 cm at the downstream face (towards downstream), see Fig.11.

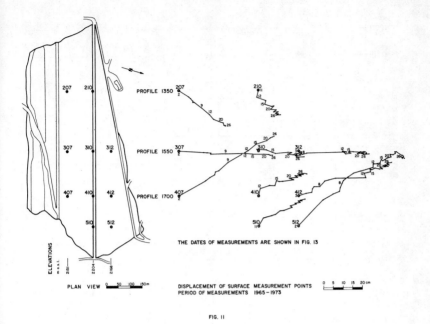

FIG. 11

The horizontal displacements of the dam base were determined in the drainage gallery. These displacements reached a maximum of 16 cm towards upstream and 2 cm towards downstream until end of 1967 (see Fig. 12)

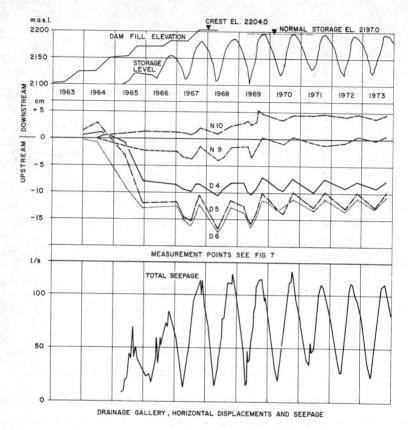

DRAINAGE GALLERY, HORIZONTAL DISPLACEMENTS AND SEEPAGE

FIG. 12

The amount of seepage during this period of the construction, corresponding to low storage levels were the following:

Year	storage level (masl)	total seepage (l/s)
1966	2155	84
1967	2180	113

By extrapolation of these values one would obtain a seepage of approximately 140 to 150 l/s at maximum storage (see Fig.12)

6. Dam and foundations behaviour during operation (1967 - 1973)

During the first six years of the operation period the dam behaviour has been very satisfactory. The foundations settlements have shown an increase of at most 25 cm during this period; the last year's increase was of about 1 cm only (see Fig. 7 and 8).

The displacements of the dam surface points, determined by means of high-precision surveys, were of at most 60 cm in the vertical and 25 cm in the horizontal dirctions during the 6 years period. The maximum displacements occurred in the crest zone (Fig. 11 and 13).

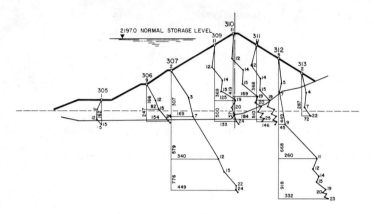

PROFILE 1550 – SURFACE MEASUREMENT POINTS

PROJECTION OF THE HORIZONTAL AND VERTICAL DISPLACEMENTS ON THE PLANE OF THE SECTION

MEASUREMENTS: MEASUREMENT NUMBER: • 24
VERTICAL DISPLACEMENT:
HORIZONTAL DISPLACEMENT: 70 mm

NR.	DATE	NR.	DATE	NR.	DATE
1	12. 7.1965	9	25. 4.1967	21	15. 10.1970
2	2. 8.	11	27. 10.	22	10. 6.1971
3	22. 9.	12	26. 4.1968	23	20. 10.
4	28. 10.	14	22. 10.	24	20. 6.1972
5	31. 1.1966	15	12. 6.1969	25	30. 10.
7	17. 6.	19	13. 12.	26	12. 6.1973
8	27. 9.	20	24. 6.1970	27	1. 11.

FIG. 13

It is interesting to point out the fact that upon depletion of the storage, the dam moves towards upstream, this motion indicating an overall elastic behaviour. The amplitude of this displacement, between the conditions of full storage and of storage completely depleted, is of about 2.5 cm.

It is also remarkable that the same motion occurs in the interior of
the dam, at the drainage gallery location. Since the amplitude of the
motion is also there 2.5 cm, it may be concluded that the dam moves
as a rigid body and that the foundation behaves as an elastic medium.

Contrarily to what was expected, the seepage losses have not increased during the operation period. The total amount of seepage at full
storage is approximately 100 l/s, from which about 65 l/s are due to
the seepage accross the foundations and about 45 l/s to the seepage
accross the core. Based upon these figures it is possible to determine the overall permeability coefficients precisely. They are $6x10^{-5}$
cm/s for the grout curtain and $3x10^{-5}$ for the core.

7. Conclusion

The Mattmark dam with a height of 120 m, a crest length of 780 m and
a total volume of 10.5 million m3 is mainly founded upon alluvial and
morainic deposits with a maximum depth of 80 m.

The treatment of the foundations was to be carried out also for reasons of economy. Due to the considerable overburden and to the occurrence of large boulders in the morainic deposits a cut-off would hard-
ly have been feasible. Therefore a treatment of the foundations by
means of grouting was carried out. The following remarks regarding the
main features of the grout curtain are of relevance:

- The number of borehole rows (10 at the top of the curtain decreasing to 4 at it's bottom) is adequate. The spacing of 3.50 m
  between these rows appears, however, to be somewhat too large.

- The clay-cement and bentonite gels used for grouting were adequate. The quantities of material used at beginning of the campaign were somewhat too high causing heave. Thus, it is suggested that grouting pressures should at most only be slightly larger than the pressure due to overburden. For a thorough grout
  penetration to be possible it is then necessary to reduce the
  speed of the operation.

The settlements which occurred in the dam foundations were considerable. The undesirable effects of such unavoidable settlements could
have been reduced either by decreasing the rate of embankment rise
or by compaction of the shell zone in layers of at most 60 cm height.

Fig. 14

Ariel view of the Mattmark dam after completion

Prior to the crack formation the rate of embankment raise was relatively high.

The quality of the core zone wether with regard to it's densitiy,
wether to it's imperviousness is excellent. The resulting k value
of 3×10^{-5} cm/s is possibly an optimal value for the material used.

The dam stability is also guaranteed by the drainage gallery which
eliminates the uplift effects under the shell zone.

After 6 years of operation the dam and it's foundations have reached
a state of quasi-elastic behaviour.

FOUNDATION OF CERRO DE ORO DAM, MEXICO

by Raúl J. Marsal* and Aurelio Benassini**

INTRODUCTION

The Secretaría de Recursos Hidráulicos (SRH) is designing the Cerro de Oro Dam, which is to be built on the Santo Domingo River, one of the main tributaries of the Papaloapan River. This dam, in conjunction with the Presidente Aleman Dam built in 1953 on the Tonto River, will contribute to the solution of a long-standing problem affecting the Papaloapan basin, i.e., the flooding of the plain downstream of the Tonto and Santo Domingo confluence.

The dam (Fig 1) is of rockfill type, 55 m high and 1,700 m long, the volume of materials being 8.7×10^6 m^3. The total capacity of the reservoir is $3,500 \times 10^6$ m^3, of which 600×10^6 are allowed for silt storage. Three tunnel spillways, with a maximum discharge capacity of 6,000 m^3/sec, are located in the left abutment; they will incorporate part of the diversion tunnels, 12 m in diameter (Fig 1).

The reservoir of Cerro de Oro Dam will be connected to that of the Presidente Aleman Dam (Fig 2) through a breach 100 m wide at the Pescaditos Dike, thus allowing an increase in the present installed power of from 154 MW to about 450 MW and also improve the utility of the system for irrigation purposes. The combined capacity of the two reservoirs will be 13.4×10^9 m^3 and flood control storage will amount to 5.2×10^9 m^3.

The erratic composition of the alluvial deposits and the permeability of the underlying rock, make the

---

*Research Professor, Instituto de Ingeniería, UNAM
**Head of Advisors' Office, Secretaría de Recursos Hidráulicos

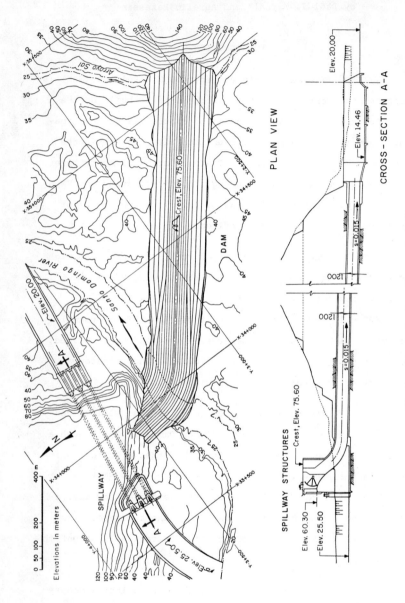

PLAN VIEW

CROSS-SECTION A-A

SPILLWAY STRUCTURES

Fig 1. Cerro de Oro Dam and Spillway

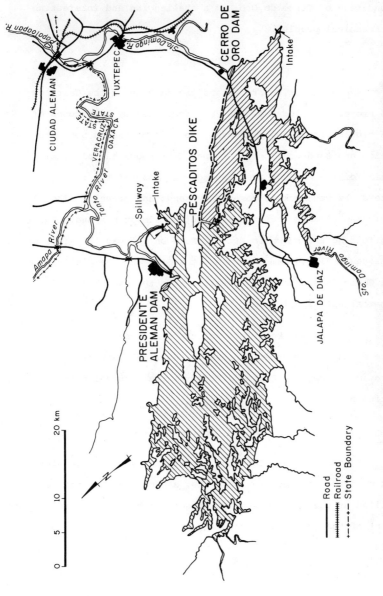

Fig 2. Reservoirs and location of Presidente Alemán and Cerro de Oro dams

foundation of Cerro de Oro Dam a challenging and interesting geotechnical problem.

GEOLOGY

The reservoir and the damsite of the Cerro de Oro Project are located in a zone dominated by limestones of the Middle Cretaceous. However, due to folding in some areas, shales and sandstones of the Upper Cretaceous and alluvial deposits of the Miocene are exposed. In most of the rivers and creeks the bed-rock is covered by sands and gravels. The tectonic map of this region and a section almost perpendicular to the folding are shown in Fig 3. Note on the map the sites of Presidente Aleman Dam and Pescaditos Dike and the placement of the Cerro de Oro damsite in relation to the Paso Nacional-Temazcal anticlinal.

The karsticity of the limestone and the presence of an old alluvial terrace in the right bank are the features of concern at Cerro de Oro from the standpoint of foundation engineering.

DAMSITE EXPLORATION

To disclose the stratigraphy, obtain samples and perform water absorption tests, a number of borings were drilled at the site in several stages. Table 1 gives an account of the exploratory work and Fig 4 shows the location of points investigated.

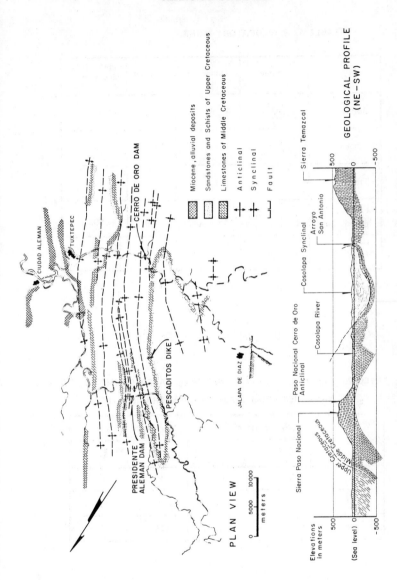

PLAN VIEW

CIUDAD ALEMAN
TUXTEPEC
CERRO DE ORO DAM
PRESIDENTE ALEMAN DAM
PESCADITOS DIKE
JALAPA DE DIAZ

Miocene, alluvial deposits
Sandstones and Schists of Upper Cretaceous
Limestones of Middle Cretaceous
Anticlinal
Synclinal
Fault

0   5000   10000
meters

GEOLOGICAL PROFILE (NE - SW)

Sierra Temazcal
Arroyo San Antonio
Cosolapa Synclinal
Cosolapa River
Paso Nacional Cerro de Oro Anticlinal
Sierra Paso Nacional

Upper Cretaceous
Middle Cretaceous

Elevations in meters
500
(Sea level) 0
-500

500
0
-500

Fig 3. Geology and tectonics of Temazcal – Cerro de Oro area

TABLE 1. EXPLORATORY WORK

Type	Name	Dimension	Number	Approx. total length (m)
Boring	P–I to P–CXVI	AX	116	4 500
Open pit	PCA–1 to PCA–10	2 x 2 m	10	100
Trench	T – 1	max. 10 x 15 m	1	150
Standard penetration	A – 1 to 7 G – 1 to 7	5 cm diam.	49	500
Addit	S – I to S – III	2 x 3 m	3	180

TABLE 2. ALLUVIAL TERRACE. NATURAL WATER CONTENTS AND N–VALUES FROM STANDARD PENETRATION TESTS

Type of soil	Nat. water content		N – values	
	Average %	Sd. deviation %	Average %	Sd. deviation %
CL	26.1	4.1	14	7
CH	39.5	8.8	19	11
ML and MH	31.7	7.7	16	7

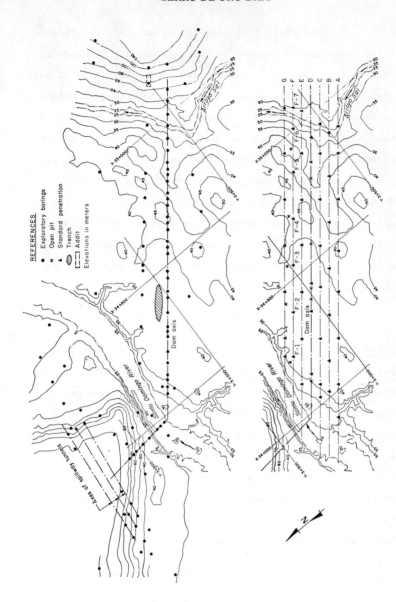

Fig 4. Location of exploratory borings, open pits, trench and standard penetration tests

The first geological campaign (1943), indicated
that the limestone is covered by a recent deposit of sands and
gravels at the river, which runs close to the left side of the
valley. On the right side, the rock is concealed by a thick,
old terrace for a distance of about 1,200 m. This terrace is
crossed by a creek (Arroyo Sal) near the right outcrop of the
limestone. Since the thickness and composition of the terrace
vary widely within the damsite, the second stage of exploration included open pits and standard penetration tests (Table 1 and Fig 4). Furthermore, a trench 20 m wide and 150 m
long excavated down to the rock permitted the direct observation of the stratigraphic sequence of the terrace as well as
of high spots of the limestone.

BASAL ROCK

Longitudinal Profile. Fig 5 presents the results of the geological exploration along the dam axis. The river channel has
a width of about 100 m where the alluvial deposit reaches a
maximum depth of 12 m. On the right bank, the limestone underlies a soil formation of variable composition and thickness, the latter ranging from 3 to 25 m. There is a section
(50 m) in the vicinity of Arroyo Sal where the rock is practically exposed; the bed materials in this creek are clean gravels and sands. The rock in both abutments above elevation 35
is covered with a layer of talus 15 to 25 m thick.

Rock Samples. Cores 2 inches in diameter, obtained with rota-

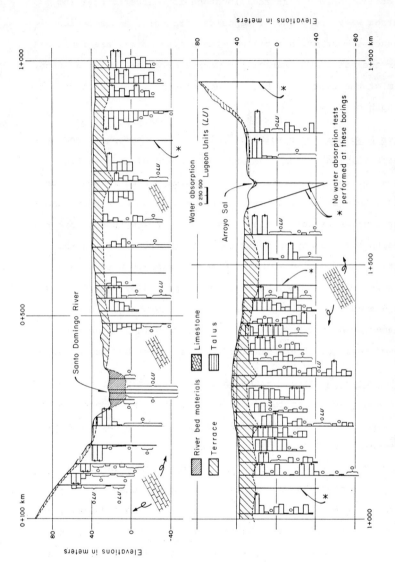

Fig 5. Damsite profile and water absorption values

ry drills, showed that the limestone is' weathered near the
contact with the alluvial deposits; fractures and solution
channels, most of them filled with clayey soils, are found all
over the site. The karsticity varies erratically in the upper
40 m of the rock mass, but in some places it may reach down to
a depth of 70 m.

Water Absorption. Results of this kind of test performed sys-
tematically in the exploratory borings are plotted in Fig 5,
values of water absorption (WA) being expressed in Lugeon u-
nits. This information reveals changes in WA with depth and
location; WA-values in the range of 100 to 400 Lugeons are not
exceptional and may be found close to the contact with the
terrace in some places and in others deeper than 50 m, e.g.,
at St. 1 + 280. Also, the section comprised by Sts. 0 + 480
and 0 + 800 show no significant water absorption at depths
greater than 30 m. Conditions of the rock in both abutments
above elevation 35 are similar to those described previously.

Rock Contact Surface. Based on the exploratory borings, open
pits and penetration tests distributed over the foundation of
the dam (Fig 4), the contour lines of the limestone below the
river-bed materials and the terrace were drawn. Fig 6 discloses that the rock surface is extremely irregular due to the
occurrence of sinks and high spots. Photographs of the exploratory trench (Fig 7) show aspects of the soil deposit, an
outcrop of the limestone and several eroded pieces of rock.
It is worth noting that during the rainy season of 1973, the

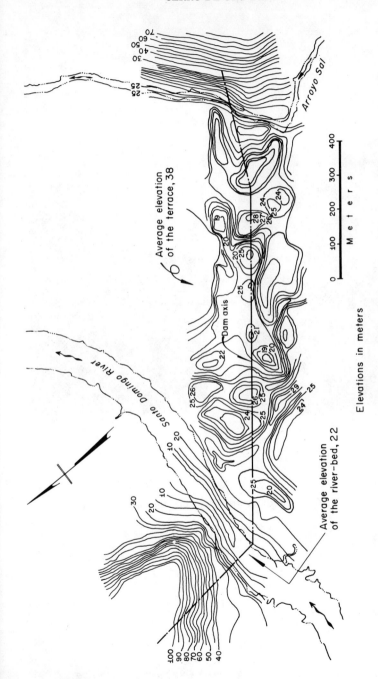

Fig 6. Contour lines of the rock surface at Cerro de Oro damsite

Fig 7. Exploratory trench in the terrace

trench was partially flooded for periods of several days, when the water in the river rose above elevation 25. Water table measurements in exploratory borings undertaken in 1973 and 1974, are presented in Fig 8. These observations reveal the critical permeability conditions of the limestone and lower boundary of the terrace which greatly influenced decisions regarding the construction of the dam and the foundation treatment.

ALLUVIAL DEPOSITS

Terrace. The old alluvial deposit on the right bank was investigated by means of 1) undisturbed samples taken at open pits and some of the borings, and 2) standard penetration tests. The latter were performed with the main purpose of detecting the types of soil and their distribution in the embankment foundation.

To illustrate the stratigraphic variations of the terrace, Fig 9 shows the longitudinal section along axis D, which approximately coincides with that adopted for the dam, and two transversal sections at Sts. 0 + 745 and 1 + 305. These profiles were simplified in order to emphasize the erraticity of the alluvial deposit. Fig 10, however, presents the detailed records of two penetration borings: C-4 and E-5. In the first, clays of both low and high plasticity, with a thickness of 16 m, cover the limestone; water contents range from 25 to 50 per cent, liquid limits are in the interval 35 to 60 per cent and plastic limits vary between 20 and

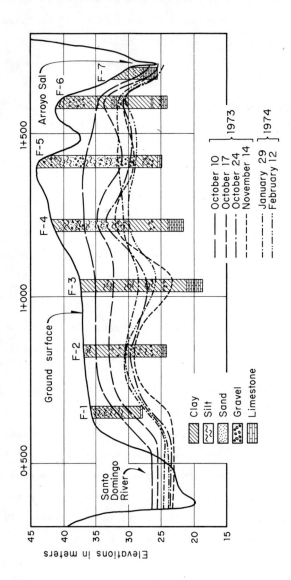

Fig 8. Water table observations at the damsite

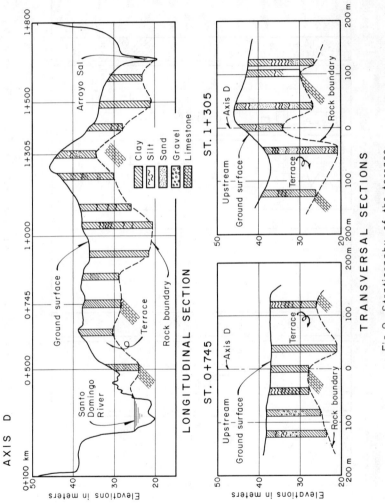

Fig 9. Stratigraphy of the terrace

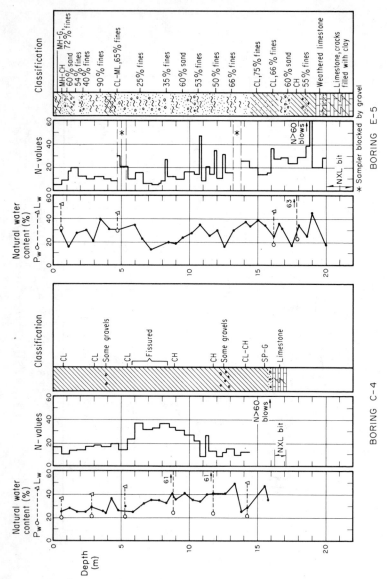

Fig 10. Results of standard penetration tests, borings C-4 and E-5

25 per cent. The blow count does not show any definite trend
with depth, N-values being greater than 10. Boring E-5, on
the other hand, discloses a sequence of soils predominantly
granular with some fines, separated by four layers of clays.
The number of blows per foot varies from 10 to 60 and water
contents are comprised between 15 and 45 per cent. To give an
overall picture of the above scattering of the data provided
by standard penetration tests, Table 2 presents average values
and standard deviations of the natural water content and number of blows N for the clays and silts tested in these exploratory borings. No correlation between the type of soil and
the blow count was established. To complement the above information, the plasticity chart of Fig 11 reveals the wide
spectrum of the cohesive soils in the terrace. In addition,
Table 3 shows the depth and thickness of the granular materials for each of the tested sites; values of both parameters
stress the erratic variation of this alluvial deposit, one of
the conspicuous characteristics to be considered in stability
analyses and prediction of settlements of the dam.

Undrained and consolidated-undrained triaxial
tests were performed with undisturbed, clayey and silty samples of the terrace (Fig 11). Most of these specimens were of
3.6 cm in diameter, a few of 10 or 15 cm, and only three of
them had a diameter of 38 cm and a height of 70 cm. Fig 12,
above, presents Mohr's envelopes that encompass those actually
obtained in both types of triaxial compression test; "average"

TABLE 3. ALLUVIAL TERRACE. DEPTH AND THICKNESS OF GRANULAR MATERIALS

Boring	Type of soil	Depth*, m	Thickness, m	Boring	Type of soil	Depth, m	Thickness, m
A–1	S–ML	1.8	3.5	E–4	–	–	–
A–2	–	–	–	E–5	SP–ML	8.8	7.3
A–3	G–M	6.0	2.0		SP	14.0	0.3
A–4	–	–	–		SP	17.1	0.3
A–5	G–ML	11.0	0.4	E–6	SP–G	2.1	0.4
	G–SP	12.9	0.2		SP–G	3.7	0.5
A–6	G–SP	4.4	0.6		SP–G	6.8	0.5
	G–SP	6.2	1.7	E–7	–	–	–
A–7	–	–	–	F–1	–	–	–
A–8	–	–	–	F–2	G–CL	10.5	0.6
B–1	–	–	–	F–3	–	–	–
B–2	G–CL	4.9	1.2	F–4	SP–G	3.2	0.4
B–3	G–SP	3.1	1.8		SP	7.0	2.0
B–4	–	–	–		SP–ML	9.5	1.0
B–5	G–CL	10.5	0.5		SP–CL	15.7	0.5
	G–SP	12.2	0.2	F–5	G–ML	5.7	0.5
B–6	G–SP	3.6	1.2		G–ML	6.8	0.4
B–7	–	–	–		SP–G	9.9	1.5
C–1	–	–	–		SP	13.4	3.8
C–2	–	–	–		G–ML	16.1	0.2
C–3	–	–	–	F–6	–	–	–
C–4	G–SP	15.8	0.4	F–7	–	–	–
C–5	–	–	–	G–1	–	–	–
C–6	SP–G	4.0	0.2	G–2	–	–	–
	SP–ML	17.0	1.9	G–3	SP–G	8.3	0.2
C–7	–	–	–	G–4	–	–	–
C–8	–	–	–	G–5	G–SP	11.5	0.5
E–1	–	–	–		S–CL	15.0	0.3
E–2	SP–G	5.6	0.30	G–6	–	–	–
E–3	–	–	–	G–7	–	–	–

* Measured from the ground surface to the middle of the stratum.

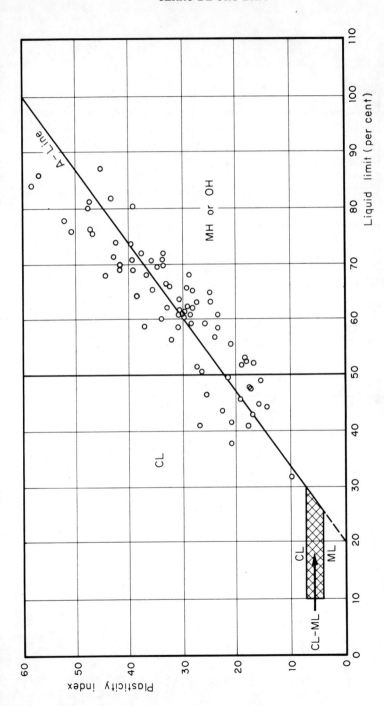

Fig 11. Plasticity chart. Clayey and silty soils of the terrace

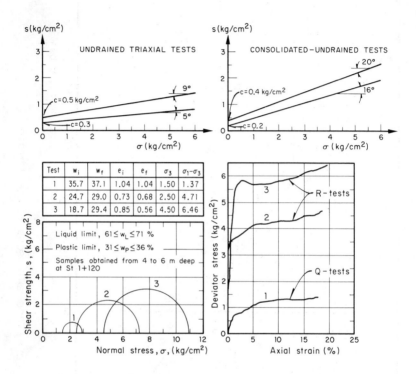

Fig 12. Shear strength of clayey and silty soils of the terrace

envelopes are also plotted. In the same figure, below, notice
the Mohr's circles at failure, pertaining test data, and de-
viator stress vs axial strain curves for the specimens 38 cm
in diameter of a clay (CH) that falls on the A-line (Fig 11).
The initial water contents and void ratios are so different
that results of these tests cannot be compared. Curves
$(\sigma_1 - \sigma_3)$ vs ε_a reveal an almost plastic behavior for $\varepsilon_a > 5\%$.

       Some of the clays and silts of the terrace were
tested in one-dimensional consolidation. Fig 13 shows void
ratio vs log. pressure and deformation vs log. time curves for
two specimens of the same clay described previously (Fig 12,
below). One of these specimens was saturated before applying
any pressure and the other at the end of the loading process.
The average coefficient of compressibility for the pressure
interval 0 to 8 kg/cm^2 is 0.01 cm^2/kg. Experimental data provided by other consolidation tests exhibit a remarkable scattering.

River-Bed Materials. Representative samples of the alluvial
deposit in the river channel were not taken. However, washed
material from the drilling indicated that it was composed by
sands and gravels; no silt or clay layers were detected.

GROUTING TESTS

       In view of the results of water absorption
tests performed in the limestone (Fig 5) and the quick response of the water levels inside the exploratory trench to

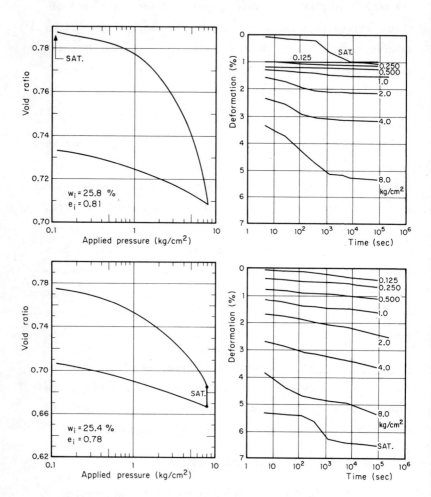

Fig 13. One-dimensional compression tests. Clay sample at
St 1+120, 5m deep

rainfall and river flows, it was foreseen that a thorough
treatment to seal the rock would be required. Estimates of
its cost were uncertain because of problems which could be anticipated both in drilling the karstic rock and grouting its
contact with the terrace. Thus the decision was taken to
perform a large scale test along the dam axis, between Sts.
1 + 335 and 1 + 435 (Fig 5). This test section was divided in
to three parts, each one being about 33 m long.

Test Section I. The layout of the first section (St. 1 + 335
to 1 + 365) included 7 lines and 46 holes distributed as shown
in Fig 14; the attached cross-section exhibits the depth of
the borings for each line. The holes through the terrace were
lined with 4 in. steel tubes, leaving a gap of about 50 cm
over the rock surface. Grouting operations were conducted as
follows: 1) the contact zone including the upper 1.5 m of
rock; 2) the karstic limestone, and 3) the underlying rock
formation to a maximum depth of 70 m, in stages. Problems developed in the drilling of limestone due to solution channels
and grout consumption was very high in the contact zone. Table 4 gives an account of the total lenght of drilling, the
amount of grout injected, the rate of grout absorption per meter, and the cost in US dollars per meter of cutoff. The exceptional consumption per meter at the contact of 212 m^3/m is
the result of dividing the grout take by the assumed 50 cm
length of injection, which is an unrealistic measure. Several
holes proved to be interconnected during grouting, and occa-

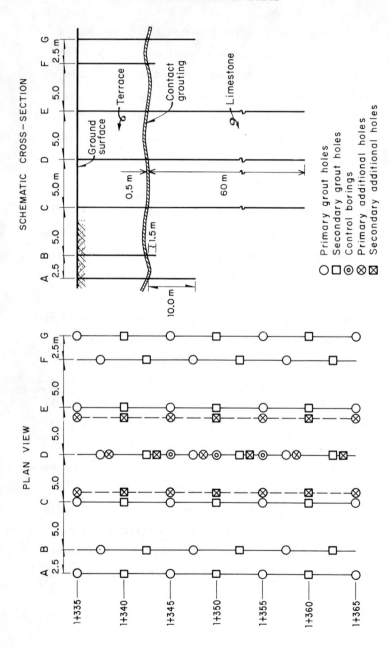

Fig 14. Grouting tests, Section I

TABLE 4. RESULTS OF THE GROUTING TESTS PERFORMED BETWEEN STS. 1 + 335 AND 1 + 435

Test section	I (St. 1+335 to 1+365)		II (St. 1+405 to 1+435)		III (St. 1+365 to 1+405)	
Formation	Terrace	Limestone	Terrace	Limestone	Terrace	Limestone
Length of drilling (m)	1 223	1 712	1 399	2 144	569	1 105
Injected grout (m³)	1 484	1 682	787	2 457	436	1 221
Grout take per injected unit length (m³/m)	212	0.98	0.83	1.15	1.65	1.11
Cost per meter of cutoff in US dollars	3 200	8 000	2 400	7 200	800	3 600

sionally the grout mixture flowed to the surface through vertical cracks, although the pressure applied was limited to 3 kg/cm^2. Note that in this test section (Fig 13), 20 holes lined with 2 in. plastic tubes sunk 1 to 2 m into the limestone were added along the three central lines in order to regrout the karstic rock.

Test Section II. The second grout section (St. 1 + 405 to 1 + 435), was planned as a 7-line treatment with a total of 79 holes. Casing through the terrace was provided by means of 2 in. plastic tubes, which were perforated (tube-à-manchettes) from 5 m deep to the rock surface. The location and depth of borings are shown in Fig 15. Using rubber packings, layers 30 cm thick of the terrace were first injected. Then, in sections 5 m long, the underlying limestone was grouted in several stages until a pressure of about 8 kg/cm^2 was attained, but limiting consumption in each operation. The data pertaining to the work done appears in Table 4. The use of plastic tubes to line the terrace materials proved to be inadequate, particularly at the rock contact. But its treatment improved noticeably with respect to that at Test Section I, and control during the subsequent drilling and grouting of the karstic limestone was good.

Test Section III. This section extends from St. 1 + 365 to 1 + 405. In a first stage 22 holes, distributed in 3 lines, were grouted; the second stage included 10 additional borings staggered with respect to the first (Fig 16). The lower por-

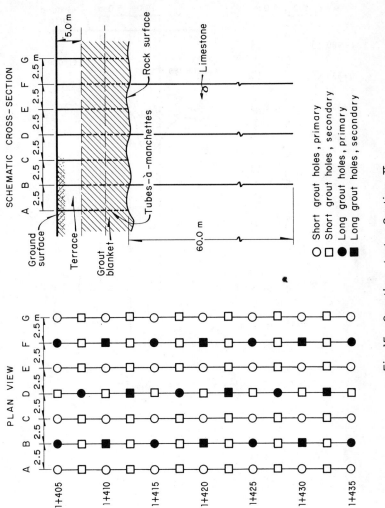

Fig 15. Grouting tests, Section II

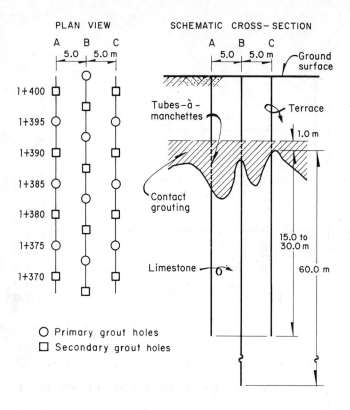

Fig 16. Grouting tests, Section III

tion of the terrace and particulary its contact with the karstic limestone, was performed by means of tubes-à-manchettes.
The rock was treated in a manner similar to that of Test Section II. For grout takes, length of drilling and cost, see
Table 4. Again, the use of plastic casing caused some problems when placing the rubber packings and, consequently, in
the grouting done at the contact zone. As disclosed by Table 4, grout absorptions are practically the same as those
registered in Section II, both for the terrace and the limestone.

Grout Mixtures and Pressures. In both the contact zone and
the karstic limestone, stabilized mixtures of water, cement,
bentonite, and sodium silicate were used. At some particular
stages where the consumption was high, fine sand was added.
The proportion of the ingredients was changed on the basis of
observed grout takes. A typical grout mixture contained
860 liters of water, 345 liters of cement, 35 kg of bentonite
and 10 liters of sodium silicated.

The maximum injection pressures were as follows:

Test Section I: contact zone, 3 kg/cm^2
                         limestone, 8 to 10 kg/cm^2

Test Sections II and III: terrace, 7 to 8 kg/cm^2
                          limestone, 10 to 15 kg/cm^2

It should be mentioned that in the latter sections, the volume of injected grout mixture was limited to
given values and the applied pressures increased from 3 kg/cm^2
to those tabulated above, in several stages.

Control Tests. A total of nine holes in the grouted sections
were drilled (Figs 14, 15 and 16) in order to perform permeability and water absorption tests, the first (Lefranc) in the
treated soils of the terrace and the second (Lugeon) in the
underlying limestone. The horizontal coefficient of permeability of the treated terrace varied over the interval 10^{-5} to
10^{-4} cm/sec and the water absorption in the rock was less than
2 Lugeons. These values apply to Test Sections II and III.
Additional grouting will be required at Section I in order to
reduce the permeability of the contact zone and the upper layers of the rock formation.

Costs. As shown in Table 4, the cost of grouting per meter of
cutoff fell in Test Section III to less than half the amount
expended on Section I. This sustantial saving reflects the
changes in grouting procedures and drilling equipment as well
as the length of boring required to attain a reasonable treatment of the contact zone and the karstic limestone. However,
to confirm the results obtained in Test Section III, which were
probably influenced by the grouting done in the adjacent Sections I and II, a treatment similar to that applied in Séc-
tion III was carried out between Sts. 1 + 135 and 1 + 235.
Lines, location and depth of borings, and the order of grout-

ing operations are shown in Fig 17. This work is currently in
progress (about 30 per cent completed); no substantial differences in grout takes as compared to those of Test Section III
were found.

DESIGN OF THE DAM

Embankment Materials. Downstream, in the vicinity of the damsite, there are large quantities of clayey soils suitable for
the impervious core as well as outcrops of limestone in both
banks adequate as sources of rockfill. Sands and gravels for
transition zones and filters were located in the river, 7 km
from the site. The cohesive soils would be the most economical material for the construction of the dam, because they are
found on both sides of the river at a distance not exceeding
2 km; their exploitation can be carried out with standard e-
quipment. But one has to take into account the fact that the
rainfall in this region is greater than 2,000 mm (80 in.) and
that in a normal year only 4 months are dry. Therefore, a dam
built mainly with clayey soil will involve control problems
with regard to the placement of this kind of material. On the
other hand, a substantial volume of rockfill will be obtained
from the excavations for the access channel, portals, tunnels
and outlet structures of the spillway.

The mechanical properties of the soils for the
impervious core and granular samples excavated from the river
borrow area, were determined in the laboratory. Table 5 pres-

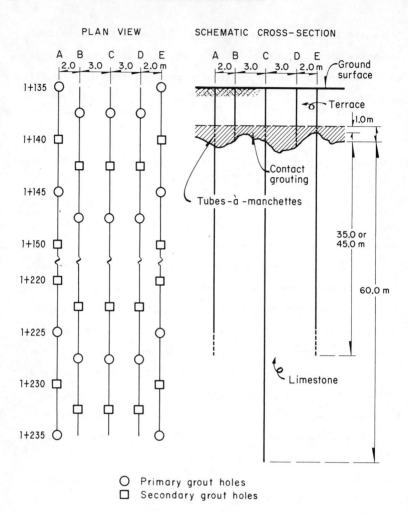

Fig 17. Additional grouting test section

ents values adopted for the design. Those of the rockfill
were selected upon the basis of test results for a similar
rock used at La Angostura Dam.

Foundation Treatment. At the site one can distinguish three
different sections as regard to the foundation of the dam:
1) the river channel, about 100 m wide, where the bed-rock underlies a thick deposit of sand and gravel; 2) the terrace on
the right bank that extends over 1,200 m, and 3) the abutments
above elevation 35, at which height the limestone is covered
by a layer of talus of variable thickness.

The main concern about this site is the permeability of the limestone due to fracturing and karsticity. This
will be solved by means of a thorough grouting of the basal
rock, in a 5-line treatment through borings at 2.5 m centers,
approximately, making full use of the experience gained in the
field tests previously described. Grouting of both abutments
will be undertaken in a similar manner, from galleries that
penetrate about 100 m inside the hill, at elevations 50 and 75.

Water table observations in the terrace (Fig 8)
indicated that serious construction problems may develop if
deep excavations are made to site the impervious core in a
trench, so as to reach the basal rock. A similar condition occurs in the river channel, because the alluvial deposit is fed
by the underlaying karstic limestone. Hence, the following
recommendations were made: 1) Along the terrace (St. 0 + 400

to 1 + 600): Build the impervious core in a trench 30 m wide
and 3 to 5 m deep, and treat the remaining soil deposit, when
required, either with a series of borings provided with tubes-
à-manchettes, or cast in place a soil-cement cutoff that penetrates into the limestone previously sealed with grout mixtures
(Fig 18). 2) In the river section (St. 0 + 300 to 0 + 400):
Place between the cofferdams a layer of sand and gravel about
5 m thick, under water, in order to raise the bottom above the
river level (dry season); under the core contact zone, this
fill will be topped with an earth embankment 5 to 6 m high to
provide a supporting "roof" for the grout treatment of both
the alluvial deposit and the granular fill (Fig 18); compact
this 5 m thick sand and gravel layer with vibratory hammers
over the zone underlying the pervious sections of the dam and
grout the bed material below the core contact; finally, remove
the earth embankment and build the core.

Construction Method. The variable composition and thickness
of the terrace suggest that a convenient way to reduce differential settlements and to improve the shear strength of the
silty and clayey soils, would be to preload the foundation of
the dam. This can be done without much interference to the
construction program, by first partially building the pervious
shoulders as shown in Fig 18, so the space required for the
impervious core is left clear to perform the grouting. After
finishing this work, which will take at least one year, the
core and the remaining portions of the shoulders are constructed in a second stage.

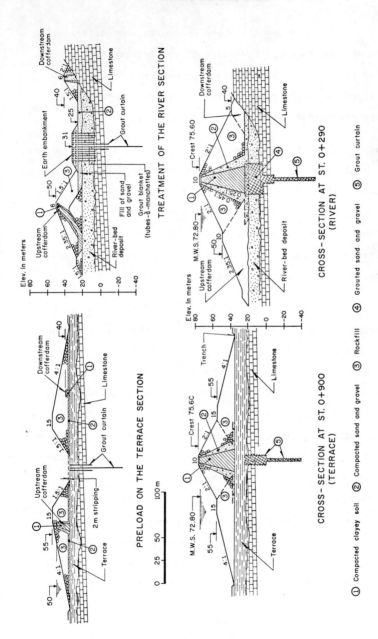

Fig 18 Excavations, preload and cross-sections of the dam

Following the above procedure, the differential settlements due to the compression of the terrace will be disposed of prior to the second stage of construction, and the shear strength of the cohesive soils will correspond to that given by consolidated-undrained triaxial tests.

Strength Parameters. The selection of parameters for the stability analysis of the dam between Sts. 0 + 400 and 1 + 600, is particularly difficult in the present case because of the different types of soil found in the terrace and their erratic distribution. On the other hand, the determination of mechanical characteristics of the materials to be used in the embankment does not pose any special problem.

If the strength parameters adopted for the terrace materials (Table 5) are compared with the values shown in Fig 12, above, one notes that the lowest plotted Mohr's envelope (c = 0.2 kg/cm^2, ϕ = 16°) obtained from consolidated-undrained triaxial tests is proposed for the cohesive soils, whereas for the granular ones c = 0 and ϕ = 30° are recommended. In addition, a potential surface of failure may intercept zones of cohesive material as well as granular lenses, since it is almost impossible to ascertain the length and type of soil along the abovementioned surface. This observation lead us to the following practical rule: use parameters c=0.2 kg/cm^2 and ϕ = 16° for the whole portion of the surface of failure located within the terrace and for stability analyses that do not include seismic forces. This is a conservative proposal

TABLE 5. MECHANICAL CHARACTERISTICS OF THE CONSTRUCTION MATERIALS FOR THE DAM AND THE TERRACE

Item	Material	Specific gravity	Compaction	Shear Strength
Impervious core	Plastic clay $L_W = 58\%$ $I_W = 27$	2.77	$w_o = 24\%$ $\gamma_d = 1620 \text{ kg/m}^3$	Undrained tests $c = 0.7 \text{ kg/cm}^2$, $\varphi = 0°$ Consol.-undrained $c = 0.4 \text{ kg/cm}^2$, $\varphi = 18°$
Transition zones	Clean, well graded sand and gravel $d_{10} = 0.3$ mm $C_u = 80$	2.74	$\gamma_d = 2150 \text{ kg/m}^3$	Drained tests $c = 0$ $\varphi = 40°$
Shoulders	Compacted rockfill	2.60	$\gamma_d = 1800 \text{ kg/m}^3$	$c = 0$, $\varphi = 40°$ *
	Dumped rockfill		$\gamma_d = 1600 \text{ kg/m}^3$	$c = 0$, $\varphi = 45°$ *
Terrace	Clayey soils	2.75	$\gamma_d = 1500 \text{ kg/m}^3$	Consol–undrained $c = 0.2 \text{ kg/cm}^2$, $\varphi = 16°$
	Sandy and silty soils	2.68	$\gamma_d = 1600 \text{ kg/m}^3$	Drained $c = 0$, $\varphi = 30°$

* Assumed values

except when the intercepted granular lenses are subjected to
a normal effective stress smaller than 0.6 kg/cm^2, a condition
which is likely to occur near the toe. For dynamic analyses
of the embankment one has to take into account the viscous increase in the shear strength.

Dam Cross-Sections. Based on the criteria and the construction method previously discussed, the cross-sections presented
in Fig 18 seem appropriate. Note that along the terrace section of the embankment (St. 0 + 600 to 1 + 400), both rockfill
zones will be placed on a sand filter 2 m thick, after stripping the ground surface.

The river section of the embankment (St. 0+300
to 0 + 400) has the two cofferdams incorporated into the rockfill zones. It will be constructed in a single stage, upon
finishing the treatment of the foundation as described earlier.

Instrumentation. To observe the settlements caused by the
preload on the terrace and their evolution with time, a rather
large number of monuments will be needed, some of them installed at the base of the dam, and the remainder on its outer
slopes and berms (Fig 18).

At least three series of piezometers, one in
the river section and the others at two different stations of
the terrace have been recommended to evaluate the effectiveness of the cutoff built by means of grouting.

The installation of series of inclinometers in three stations close to the piezometer sections, with their casings fixed to the basal rock, will permit the measurement of horizontal and vertical displacements within the dam as well as in the river-bed materials and the terrace.

Several concrete tubes connected to the downstream trench of the embankment between Sts. 0 + 600 and 1 + 400, provided with wiers at their discharge ends, are required to gage the rate of seepage and its distribution along this section of the dam.

REMARKS

The Cerro de Oro Dam presents several problems which deserve to be emphasized.

1) The karsticity and fracturing of the basal rock, a limestone of the Middle Cretaceous, will demand a careful and expensive campaign of grouting. At present the cost is estimated at 9 million US dollars, about 12 per cent of the total cost of the project. The treatment includes the alluvial deposit in the river channel, the soils of the terrace on the right bank, and the basal rock and both abutments. In such a difficult and complex problem the experience derived from the field tests performed, has been of great help in the planning and evaluation of the work needed to reduce the risk of the serious damage which could be caused by uncontrolled seepage through the foundation of the dam.

2) The hazards involved in making excavations below certain elevations both in the river-bed material and the terrace, because of the high permeability of the basal rock, has favored a solution that avoids the handling of uncertain and large volumes of seepage during construction. This alternative calls for thorough grouting to substantially reduce the permeability and compressibility of these deposits. Also a sand and gravel fill about 5 m thick in the river section between the cofferdams, must be carefully compacted with vibratory hammers or similar.

3) The presence of the soil terrace on the right bank, however beneficial from the standpoint of seepage, has brought other problems due its variability in composition and thickness. This, added to the unfavorable condition commented upon in Point 2, has suggested the preloading of the terrace as a means to reduce differential settlements and increase the shear strength of the cohesive soils. Since, on the other hand, it is not possible to reliably estimate the length and location of the various soils in the terrace along a potential surface of failure, the proposal has been made to use the lower boundary of the band that contains the Mohr's envelopes corresponding to consolidated-undrained triaxial tests.

4) The unusual features of the Cerro de Oro Project demand a proper instrumentation of the dam, to identify unforeseen problems and to gain experience in difficult foundations of this type.

MANICOUAGAN 3 CUT-OFF

BY

YVES PIGEON*

S U M M A R Y

The Manicouagan 3 Hydro Power Development, presently under construction, features a 350 foot high earthdam resting on a 400 foot deep sediment filled canyon. Seepage control through the foundation is a-chieved by means of a vertical cut-off consisting of two parallel concrete walls made of cast in place interlocking piles and panels topped with a gallery. Construction of this part of the work was completed in 1973 and even the foundation gallery is buried in the embankment.

After describing the geological conditions at the site, design considerations are reviewed briefly before discussing construction problems and the techniques developed to solve them. A preliminary assessment of the cut-off efficiency is made and deformations of the walls to date reported. Although a definitive evaluation of the cut-off cannot be made until filling of the reservoir, it is believed that the cut-off fulfils its purpose.

INTRODUCTION

The Manicouagan 3 Hydro Power Development is located on the Manicouagan River some 55 miles north of its confluence with the Saint-Lawrence River near Baie-Comeau, Quebec, and 350 miles north-east of

---

* Head, Soil Mechanics Department
Asselin, Benoît, Boucher, Ducharme, Lapointe Inc., Consultants
Montreal, Canada.

Montreal, Quebec (Fig. 1). As part of the development (Fig. 2) a 350 foot high earthdam is being built across the river valley. The site of this dam presented a challenging problem to the designer due to the presence of nearly 400 feet of sediments in the riverbed.

The dam design finally adopted after lengthy studies includes the following main features (Fig. 3):

- A main cut-off located on the axis of the dam and consisting of two cast in place concrete walls 2 feet thick at 10 foot centers extending to bedrock. This cut-off is topped with an inspection gallery.

- A partial cut-off at the upstream cofferdam linked by a horizontal blanket to the core of the dam. A horizontal drainage layer at the base of the downstream shell completes this second group of seepage control measures.

- A fairly thick central core earth embankment with average outside slopes of 3:1.

Design considerations both for the embankment and the foundation treatment have already been discussed in a number of technical papers. After reviewing geological conditions, these design considerations will be summarized briefly before turning to construction experience and a preliminary assessment of the performance of the cut-off.

GEOLOGICAL CONDITIONS

The existence of a sediment filled canyon below the riverbed was recognized at the time when the dam site was chosen. However, its depth was not known, nor the characteristics of the sediments. Investigations

(1) Manicouagan 2.
(2) Manicouagan 3.
(3) Manicouagan 5.
(4) Rivière Manicouagan river.

FIG. I CARTE DE LA PROVINCE
MAP OF THE PROVINCE

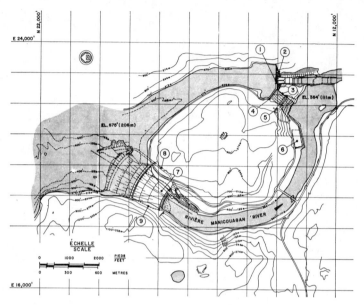

AGENCEMENT GÉNÉRAL - GENERAL ARRANGEMENT

FIG. 2 **AMÉNAGEMENT HYDRO-ÉLECTRIQUE MANICOUAGAN 3**
MANICOUAGAN 3 HYDROELECTRIC DEVELOPMENT

① Déversoir
 Spillway

② Barrage-poids
 Gravity dam

③ Passe à billes
 Log chute

④ Prises d'eau
 Intakes

⑤ Usine souterraine
 Underground power plant

⑥ Poste de sectionnement
 Switchyard

⑦ Barrage principal
 Main dam

⑧ Galerie de dérivation
 Diversion tunnel

⑨ Accès galerie de fondation
 Foundation gallery access

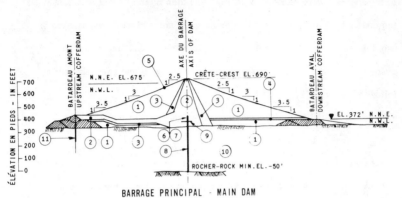

BARRAGE PRINCIPAL - MAIN DAM

GALERIE DE FONDATION   FOUNDATION GALLERY

(1) GRANULAIRE COMPACTÉ
COMPACTED GRANULAR MATERIAL

(2) TILL COMPACTÉ
COMPACTED TILL

(3) TRANSITION
TRANSITION

(4) DRAIN
DRAIN

(5) PERRÉ
RIPRAP

(6) ZONE DE BENTONITE
BENTONITE ZONE

(7) GALERIE DE FONDATION
FOUNDATION GALLERY

(8) DIAPHRAGME D'ÉTANCHÉITÉ
MAIN CUT-OFF

(9) DIAPHRAGME D'ACIER
STEEL DIAPHRAGM

(10) ALLUVIONS
ALLUVIUMS

(11) COUPURE PARTIELLE
PARTIAL CUT-OFF

BARRAGE PRINCIPAL ET GALERIE DE FONDATION
COUPES TYPES

FIG. 3 ────────────────────────────

MAIN DAM AND FOUNDATION GALLERY
TYPICAL SECTIONS

in the riverbed were carried out in 1957 and then intermittently between
the end of 1963 and 1969 when the decision to proceed with construction
was taken. Some 129 boreholes (Fig. 4) totaling 15,330 linear feet in
overburden and 12,920 linear feet in rock were drilled and samples by the
following methods:

- Percussion drilling with standard penetration and permeability tests.
  All of these holes were cased.
- Rotary drilling with mud in uncased holes to obtain undisturbed samples
  below the water table. A thin walled piston sampler as recommended
  by the Waterways Experiment Station was used for this purpose.
- Rotary core drilling in bedrock.

From the borehole data, bedrock contours were traced over the dam
site and the approximate shape of the canyon determined. The cross-
section of Fig. 5 shows a typical "U" shaped valley of glacial origin
deepened by a "V" gorge extending approximately from El. 175' to El. -50'.
The bedrock is a coarse grained anorthosite of precambrian age intersected
by numerous basic dykes. Three groups of widely spaced joints were
mapped in this area: one is horizontal and most likely associated with
stress relief during the retreat of glaciers; the other two are nearly vertical
and strike in the N-S and E-O directions. Weathering of the exposed rock
surfaces is not important. Unconfined compression tests on rock cores
indicate an average strength of about 13,000 psi for submerged samples.

Sediments in the valley are of fluvial and glacial origin and vary
from fine silty sands to boulders rarely exceeding 3 feet in diameter.

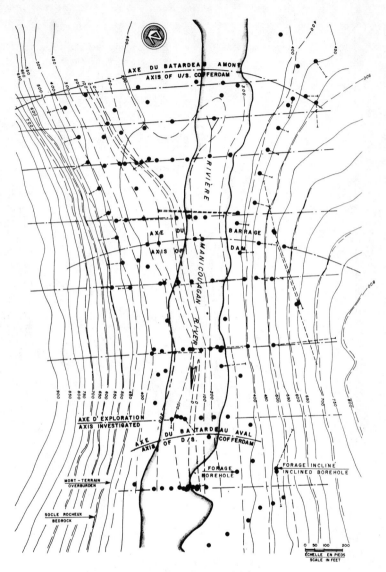

FIG. 4 EXPLORATION DES FONDATIONS
        FOUNDATION INVESTIGATIONS

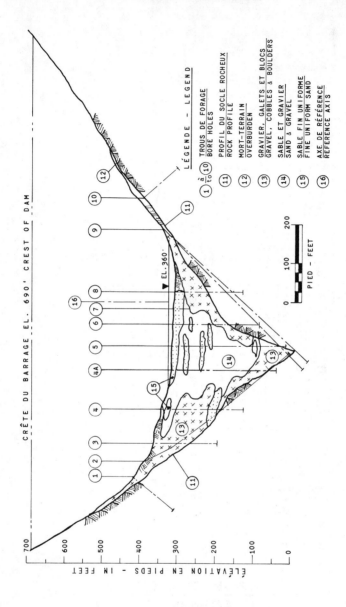

FIG. 5　　COUPE TYPE DU SILLON
TYPICAL CANYON CROSS-SECTION

LÉGENDE – LEGEND

(1 à 10) TROUS DE FORAGE
(1 to 10) BORE HOLES

(11) PROFIL DU SOCLE ROCHEUX
ROCK PROFILE

(12) MORT-TERRAIN
OVERBURDEN

(13) GRAVIER, GALETS ET BLOCS
GRAVEL, COBBLES & BOULDERS

(14) SABLE ET GRAVIER
SAND & GRAVEL

(15) SABLE FIN UNIFORME
FINE UNIFORM SAND

(16) AXE DE RÉFÉRENCE
REFERENCE AXIS

CRÊTE DU BARRAGE EL. 690' CREST OF DAM

ÉLÉVATION EN PIEDS – IN FEET

PIED – FEET

Generally, the boulders are located in the gorge and near bedrock with the sands and gravels in the center of the valley.

Penetration Resistance

As would be expected in such sediments, the penetration resistance varies from a few blows per foot to refusal with no particular pattern except for a definite increase in the minimum number of blows with depth (Fig. 6).

Permeability

Most of the permeability tests were performed inside the casing and are thus punctual tests of limited accuracy. Measured values ranged between 10^{-5} cm/s to 10^0 cm/s with no apparent pattern or layering. However, the stratigraphy of the sediments in the canyon did indicate the presence of coarse material near the bottom and extensive zones of fine sand below the riverbed and on the river terrace on the right bank at the upstream cofferdam location. This represents the extent of the information available when the foundation treatment was designed, but subsequently full scale pumping to dewater the dam foundation (Figs 14 to 16) to riverbed elevation provided more reliable information on this subject.

Initially, to design the dewatering system, an average permeability of 5×10^{-2} cm/s was assumed and subsequently confirmed by a pumping test in a well 250 feet deep. The 14 wells installed afterwards at the upstream and downstream cofferdams showed that this permeability was valid below about El. 250' but not in the upper part of the canyon. Wells with screens above this elevation had a much lower specific capacity.

Thus, the canyon cross-section can be divided into an upper layer

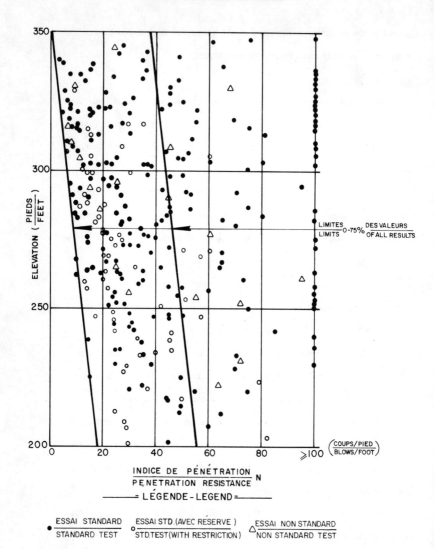

FIG. 6   PROFIL DE L'INDICE DE PÉNÉTRATION STANDARD
         PROFILE OF STANDARD PENETRATION RESISTANCE

of fine and silty material down to El. 250' followed by sand and gravel with cobbles and boulders having an average permeability of 5×10^{-2} cm/s.

Fine Sand Characteristics

Approximately 50 undisturbed samples of fine sand (Fig. 7) were recovered from 4 boreholes out of 9 of this type which were drilled below the riverbed during the exploration program. The purpose of this sampling was to enable the designer to assess the liquefaction potential of the fine sands during an earthquake. Standard classification tests on 39 samples gave the following average values :

  unit weight   : 103 lb./ft^3

  uniformity coefficient :   3

  relative density   :   53%

Triaxial tests on two samples gave a minimum shearing resistance of 35° at a relative density of 10%. This result is confirmed by the low roundness and the angularity of the individual grains inspected under a microscope.

FOUNDATION TREATMENT

Design Considerations

The large depth of overburden beneath the riverbed made it uneconomical to excavate to bedrock and thus ruled out the possibility of a concrete dam. Consequently, studies were centered on earth embankments and a central core was adopted in view of the poor characteristics of the available impervious material. Foundation treatment for such an embankment has to provide for the following :

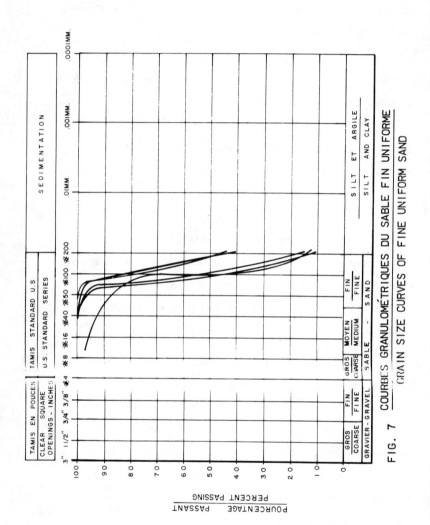

FIG. 7   COURBES GRANULOMÉTRIQUES DU SABLE FIN UNIFORME
GRAIN SIZE CURVES OF FINE UNIFORM SAND

a)   Ensure that seepage through the foundation is sufficiently small to
        avoid a significant loss of water.

b)   Control of that water which does seep through the foundation to a-
        void internal erosion.

c)   Control of pore pressure in the downstream half of the dam to en-
        sure the stability of the embankment.

Having ruled out excavation to bedrock, two other possibilities
were considered for foundation treatment: an impervious upstream blanket
and a vertical cut-off. However, the former scheme had to be abandoned
due to the difficulties involved in the construction of an impervious blanket
all over the terraces on the river-banks and to the lack of suitable impervious materials within a reasonable distance. Moreover, this scheme implied a lengthening of the 55 foot diameter diversion tunnel with conse-
quent increase in cost. Hence the decision to build a vertical cut-off.

The main design requirement for the cut-off is the creation of an
effective barrier against water seepage which will maintain its efficiency
under severe loading conditions and with time. Consideration was given
to the construction of a grout curtain. However, the presence in the foundation of large zones of fine sands which would be difficult to grout, the
danger of severe cracking within the grouted mass under large static and
dynamic loads and the difficulty of ensuring completion within a tight
schedule made this scheme less attractive than a positive cut-off consisting of two parallel concrete walls of the Icos type topped with a gallery.
Full scale pile tests carried out on site in 1968-1969 proved that such

piles could be built to a depth somewhat in excess of 400 feet.

The twin walls serve to reduce the load applied on each wall as the sediments around it settle under the weight of the embankment and to allow the construction of a gallery at the foundation level. This gallery provides access to monitor the behaviour of the cut-off walls and to undertake remedial grouting between the walls if necessary.

In view of the unprecedented nature of this cut-off with regard to depth and loading conditions, advantage was taken of construction needs to incorporate additional seepage control measures. The upstream cofferdam design includes a partial cut-off made of 2 foot thick concrete panels which is linked to the core through a 30 foot thick impervious blanket (Fig. 3). This system coupled with a horizontal drain near the base of the downstream shell is sufficient to ensure the stability of the dam in the event of a serious malfunction of the main cut-off.

Description of the Cut-off

The cut-off consists of 2 foot thick parallel concrete walls spaced 10 feet apart from center to center (Fig. 8). They are located symmetrically about the axis of the dam which is curved on a 2,000 foot radius. A minimum key of 2 feet in bedrock has been provided everywhere at the bottom of the walls. Where the depth to bedrock is in excess of about 170 feet, the walls are made of interlocking piles. Elsewhere panels having a length of 11 feet are used. Cross-walls located on both sides of the canyon separate the shallow and the deep sections. The specified concrete strength is 5,000 psi. The total surface area of each wall in-

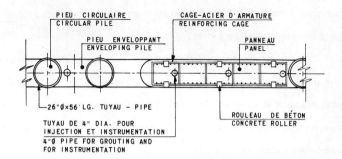

PIEU CIRCULAIRE
CIRCULAR PILE

PIEU ENVELOPPANT
ENVELOPING PILE

CAGE-ACIER D'ARMATURE
REINFORCING CAGE

PANNEAU
PANEL

26"Ø×56'LG. TUYAU – PIPE

TUYAU DE 4" DIA. POUR
INJECTION ET INSTRUMENTATION
4"Ø PIPE FOR GROUTING AND
FOR INSTRUMENTATION

ROULEAU DE BÉTON
CONCRETE ROLLER

PIEUX ET PANNEAU TYPES (PLAN PARTIEL)
TYPICAL PILE AND PANEL (PARTIAL PLAN)

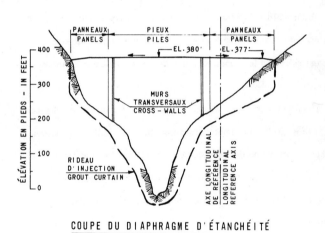

PANNEAUX
PANELS

PIEUX
PILES

PANNEAUX
PANELS

EL·380'   EL·377'

ÉLÉVATION EN PIEDS – IN FEET

400

300

MURS
TRANSVERSAUX
CROSS – WALLS

200

100

RIDEAU
D'INJECTION
GROUT CURTAIN

AXE LONGITUDINAL
DE RÉFÉRENCE
LONGITUDINAL
REFERENCE AXIS

0

COUPE DU DIAPHRAGME D'ÉTANCHÉITÉ
CUT - OFF SECTION

FIG. 8   DIAPHRAGME D'ÉTANCHÉITÉ - PLAN ET COUPE
CUT-OFF - PLAN & SECTION

cluding the 2 foot key in bedrock is 115,748 square feet of which 31,367

square feet are covered by panels.

The cut-off is extended into bedrock by contact and curtain grout-

ing at the base of the walls. Four (4) inch casings were left in all enve-

loping piles and at an average spacing of 4 feet in the panels for this

purpose. Contact grouting was generally carried out at the bottom of each

casing while curtain grouting was limited to the upstream wall in zones

where water pressure testing in exploratory holes indicated a permeability

in excess of 10^{-3} cm/s.

At its upper end the cut-off penetrates 40 feet into the core to pro-

vide an adequate percolation path. In view of the high stress concen-

tration which will tend to favour cracking in the core around the walls and

the gallery, this seepage path is lengthened by a horizontal steel dia-

phragm extending 50 feet upstream and 30 feet downstream of the cut-off

at elevation 379'.

The diaphragms are made of 4 layers of 18 gauge steel sheets

coated with asphalt. For imperviousness, they are built like roofing,

each sheet overlapping the previous one by 75%. These membranes are

attached to the outside steel plate of the foundation gallery by a sliding

joint (Fig. 3) designed to accommodate the important foundation settle-

ment anticipated.

The concrete walls are topped with a reinforced concrete slab on

which rests the foundation gallery which is a two hinged ogival arch of

composite steel and concrete design. It is 8 feet wide and about 10 feet high with walls 2 feet thick so that its outside faces are flush with the concrete walls. Rock tunnels in both abutments provide access to the gallery.

To prevent high stress concentration on top of the gallery which would add to the already large loads to be supported by the walls, a zone of pure bentonite having a maximum thickness of 20 feet and 10 feet wide is placed above it at a moisture content somewhat above the plastic limit. The compressibility of the bentonite should be sufficient to prevent high stress concentrations. However, as an additional control measure, 6 inch bleeding pipes have been placed at about 7 foot intervals in the roof of the gallery to permit the removal of bentonite if stresses exceed the design values.

Instrumentation in the walls includes hydraulic and electrical piezometers, mechanical and electrical extensometers, inclinometers and shear strip failure locators. The outside of the gallery is equipped with hydraulic and electrical piezometers and total stress cells. All instruments are read from the foundation gallery or the control room at the end of the access tunnel on the left bank.

CONSTRUCTION EXPERIENCE

Work on the cut-off construction started in June 1971 or approximately at the same time as work on the temporary diversion tunnel. A working platform was built from the right bank as far as possible in the riverbed without unduly restricting the riverflow. This platform ended with a 70 foot diameter cellular cofferdam which gave access in this first stage of construction to nearly all of the pile area plus the panels on the right bank. A shelter over the whole pile area enabled work to proceed throughout the winter. Completion of the walls including grouting was about one month late on an 18 month schedule and this delay was made good during the winter work (1972-1973) on the construction of the pile cap and gallery so that the working platform excavation started on time in the early spring of 1973.

This is altogether a quite satisfactory result considering the innovative character of the work and the problems which had to be solved during construction. The more important ones of these will be discussed below.

Panel Construction

Where depth to bedrock did not exceed about 170 feet, the walls were made of cast in place panels. This choice was made strictly for economic reasons, the estimated cost of panels being less than half of that for piles covering the same area. However, the sediments at Manic 3 proved very hard to penetrate with a standard bucket type excavator. On the right bank, pilot holes had to be drilled by percussion almost everywhere to make any progress at all and finally a new 6 ton mechanical bucket was designed and built to complete the job.

To make matters even worse, a zone of frozen ground was encountered near the abutment between elevations 365' and 340' for a length of about 50 feet. Permafrost at this low latitude (50°) had not been reported before but the presence of this zone was confirmed during the excavation of the working platform. It could be explained by the fact that very little sun reaches this area due to the vicinity of the steep abutment and the orientation of the river valley at the dam site.

On the left bank panels were excavated with two hydraulic buckets guided by telescoping structural steel sections. Even then progress was slow and two percussion rigs had to be constantly available to crush boulders or loosen the denser material.

The rates of progress excluding downtime are given in Table I.

TABLE I

Depth	Rate of progress sq ft/hour	
	right bank	left bank
0 - 100'	7.2	14.3
100' - rock	3.0	4.4
Rock - key	0.45	0.69

The 2 foot key in bedrock which was specified also proved difficult to achieve. Due to the hardness of the rock and the steepness of the bedrock profile in some areas, the free falling chisel kept sliding on the surface and did little useful work. The Contractor solved this problem by using a steel guide for his chisel to prevent the slipping and a reverse circulation rig to remove cuttings and always provide a clean surface for the action of the chisel.

Excavation of the working platform over a height of 40 feet allowed the inspection of part of the walls. The following observations were made:
- Bulges of concrete 1 to 4 feet in excess of the average panel line were

observed mainly in the top part of the panels. Bulges in excess of 1 foot were trimmed off. Excess concrete over the theoretical volume was about 16%, but only a small part of this excess was in the form of bulges.

- Joints between panels appeared tight except for 2 which had to be cleaned-up and backfilled. One which extended below the level of excavation was grouted with cement and chemicals. Five holes to bedrock were drilled for this purpose.

- Poor concrete was always present on the sides of each panel in the top 3 or 4 feet. Such concrete was excavated and backfilled at the same time as the pile cap was poured.

- The contact between concrete and bedrock seemed good, except for one panel where poor concrete was excavated and replaced.

Pile Construction

Difficulties related to pile construction were more numerous than those associated with panels and this is not surprising in view of the unprecedented depth of the piles. The more important ones will be discussed in the following paragraphs.

Verticality of Primary Piles

The full scale pile tests performed on site in 1968-1969 indicated that piles more than 400 feet deep could be built in place. At the time of these tests, flat chisels were used by Icanda to excavate the primary piles. The circular cross-section was obtained by slowly rotating the chisels. However, deviations from the vertical in excess of the 6 inch tolerance specified occurred and had to be rectified. This implied back-filling with lean concrete up to where the deviation started and re-drilling. In addition, control of verticality was performed with a plumb-bob intro-duced in the hole after removal of the drilling rods. This is a time con-suming procedure but deviations have to be limited to ensure that the enveloping elements do reach the adjoining primary piles.

After being awarded the contract, Icanda designed and built a new system to overcome this problem. The basic drilling rig remained the same as before, but two concentric chisels are used. The pilot chisel is attached to 8 inch diameter flush jointed casing which serves as a guide for the 24 inch chisel. Inside the 8 inch casing, next to the chisel, there is a knob to which a cable can be hooked (Fig. 9). This cable after tensioning is aligned vertically and this provides a direct measurement of deviation between the top of the hole and the chisel. Since measurements can be made without removing the drill rods, verticality can be determined much more often and deviations corrected before they become excessive.

At Manic 3, readings were taken at 20 foot intervals and only 1% of 1,814 measurements exceeded the specified maximum deviation. More-over, deviations were not continuous over the whole pile (Fig. 10). Out of 152 primary piles, only 2 had to be partially backfilled with lean con-crete to correct a deviation. As already mentioned, part of this success is due to the closer check on verticality but the opportunity of using the concentric chisels as guides for one another helped considerably to pre-vent and correct large deviations.

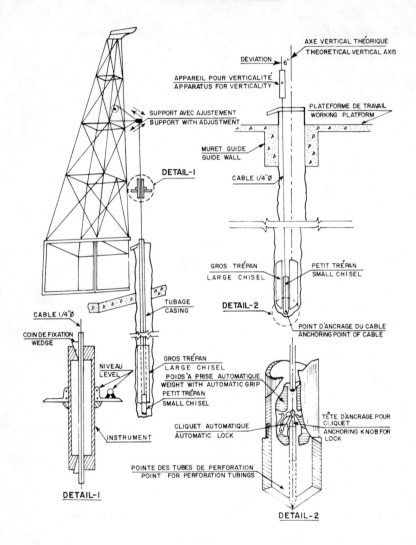

FIG. 9   MÉTHODE DE MESURE DE LA VERTICALITÉ DES PIEUX
METHOD FOR MEASURING THE VERTICAL ALIGNMENT OF PILES

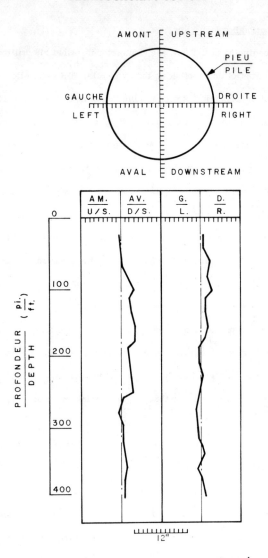

FIG. 10 MESURE TYPIQUE DE VERTICALITÉ
TYPICAL VERTICALITY MEASUREMENT

Mud Communication Between Piles

At the beginning of the job, the Contractor installed his drilling rigs directly in front of one another on the two walls. This was allowed by the technical specifications which required a minimum spacing of 10 feet between simultaneous excavations. Such a spacing had been used successfully at Manic 5 where a similar single wall cut-off had been built at the upstream cofferdam. However, it was soon discovered that where drilling took place through coarser zones of sediments, mainly gravel, cobbles and boulders, mud communications occurred between piles. This had serious effects on production rates since work had to be interrupted on one of the communicating piles until completion of the other one. Moreover, the possibility of holes collapsing could not be underestimated although in practice no such occurrences were recorded.

To solve this problem, the spacing between excavations was increased gradually up to 18 feet from center to center, at which distance no communications were noted. Communications had occurred both along the same wall and from wall to wall so that no preferential direction could be established.

While the spacing was being increased, the 24 inch chisel was modified to provide a larger open area for the mud flow. It had been speculated that the falling chisel acted as a piston which forced the mud out of a hole and into another one when traversing very coarse layers. Since this change was made at the same time as the spacing was increased, it is not possible to state with certainty that this was part of the problem, but the larger openings did allow the removal of larger cuttings resulting in improved efficiency.

As work progressed and possibly as the space between walls filled up with bentonite, this problem nearly disappeared and only occasional communications were noted at a maximum spacing of 12 feet, except in one instance where the distance was 22 feet along the same wall with completed piles in between.

Steel Casing Installation

All primary piles were reinforced with 26 inch casings $\frac{1}{2}$ inch thick and 56 feet long to withstand stresses during the excavation of the working platform on the outside of the walls. The Contractor initially tried to install these casings in holes drilled with a 26 inch diameter chisel, enlarged this to 27 inches and finally had to standardize on 28 inch chisels before a successful operation could be obtained. Over 400 machine-hours were thus lost on what is a fairly simple problem, but since the basis of the method of excavation used is hole stabilization with mud, the personnel had no experience in the installation of casings which even for the 28 inch chisel had to be hammered into place.

Rate of Progress

Average rates of drilling have been calculated for primary and enveloping piles. They are summarized in Table II.

TABLE II

Depth	Rate of progress − Ft/hour	
	Primary piles	Enveloping piles
0 − 100	1.80	2.41
100 − 200	1.47	1.19
200 − 300	0.86	0.74
300 − 400	0.63	0.68
400 − rock	0.31	0.37
Rock − key	0.44	0.32

These rates account only for actual drilling operations and exclude machine breakdown, power failures, etc. They should be multiplied by two for comparison with rates for the panels, since piles have a nominal diameter of 2 feet. Considering actual drilling time only, there should be no particular reason for the rate of progress to decrease substantially with depth, except when changes in the nature of the excavated material occur. Thus it can be seen that coarser and denser material was found as the holes went deeper.

It is also interesting to note that rates of penetration are not very different for primary and enveloping piles even though the area of the enveloping piles is much larger than that of the primary piles. The rates for enveloping piles also include the use of the expansible chisel to remove all material on adjacent primary piles.

Appearance of Walls

Inspection of the walls after excavation of the working platform revealed apparently tight joints everywhere with a layer of bentonite varying in thickness between 1/8 and 1/2 inch. No bulges of concrete of the

kind described for the panels were encountered. The volume of concrete

actually placed exceeded the theoretical amount by 22% for the primary

piles but was about 8% less than estimated for the enveloping piles.

Construction of Pile Cap and Foundation Gallery

Construction of the pile cap and foundation gallery took place in

the winter and spring of 1973. No special problems were encountered and

it was possible to make good the time lost on the wall construction and

fall back on the planned schedule of works.

PERFORMANCE OF THE CUT-OFF

Since this cut-off was the first one to be built to such a depth,

the owner, the Quebec Hydro-electric Commission, agreed that it should

be extensively instrumented to monitor its behaviour. Instruments install-

ed are as follows:

Upstream wall   :   9 vibrating wire piezometers on the upstream
(Fig. 11)   face

165 electrical extensometers

3 benchmarks with dial extensometer

3 inclinometer casings

10 shear strip failure locators

Downstream wall   :   30 vibrating wire piezometers on the upstream
(Fig. 12)   face

9 vibrating wire piezometers on the downstream
face

5 hydraulic piezometers on the upstream face

54 electrical extensometers

1 benchmark with dial extensometer

5 inclinometer casings

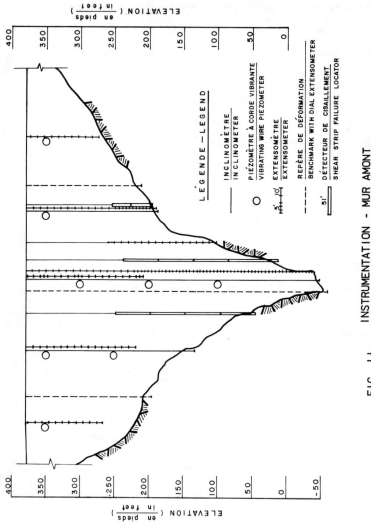

LÉGENDE — LEGEND

INCLINOMÈTRE
INCLINOMETER

PIÉZOMÈTRE À CORDE VIBRANTE
VIBRATING WIRE PIEZOMETER

EXTENSOMÈTRE
EXTENSOMETER

REPÈRE DE DÉFORMATION
BENCHMARK WITH DIAL EXTENSOMETER

DÉTECTEUR DE CISAILLEMENT
SHEAR STRIP FAILURE LOCATOR

INSTRUMENTATION - MUR AMONT
INSTRUMENTATION - UPSTREAM WALL

FIG. 11

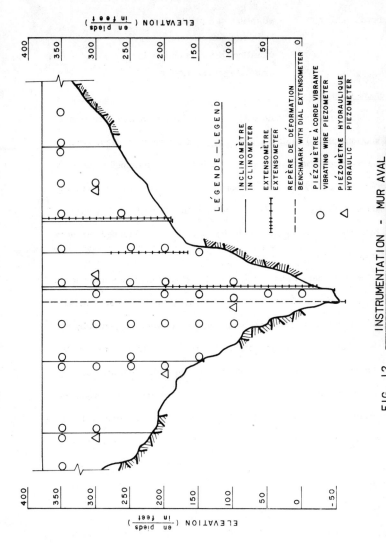

FIG. 12 INSTRUMENTATION - MUR AVAL
INSTRUMENTATION - DOWNSTREAM WALL

Foundation gallery   :   vibrating wire piezometers

                              hydraulic piezometers

                              vibrating wire pressure cells

                              hydraulic pressure cells

Instrument Installation

    Piezometers were installed inside 3 x 3 inches hollow structural

steel sections placed against the face of piles before concreting. These

sections were perforated with shaped charges at predetermined elevations

just prior to installation of the instruments to put them in contact with the

sediments. This operation was quite successful although smaller charges

could have been used. Out of 50 electrical piezometers planned, 48 were

installed in this way.

    For electrical extensometers, steel casings for contact and curtain

grouting had been replaced by P.V.C. casings. However, most of these

were damaged during installation, even though they were supported by

steel cages, and a considerable amount of drilling had to be performed to

regain the necessary 4 inch holes for installation of the instruments.

When some of the holes could not be recuperated, attempts were made to

drill from the top of adjacent piles but the maximum depth which could be

reached before holes deviated outside a pile was about 200 feet. In the

end, extensometers planned for lost P.V.C. casings were installed inside

4 inch steel casings cut at short intervals with a special tool so that the

sensitivity of the instruments would not be lost.

    Installation of other instruments presented no special problems.

Presently all instruments are being read manually from junction boxes in the foundation gallery, but all those with electrical output will soon be connected to an automatic control system located at the end of the access tunnel to the gallery on the left bank.

Piezometer Readings

The vibrating wire piezometers in the cut-off walls have a range of more than 400 psi with an accuracy of $\pm$ 3 feet. At the present time, the hydraulic head on the cut-off is about 26 feet and kept constant by control gates at the entrance to the diversion tunnel. In view of the accuracy of the instruments, only broad outlines of pressure distribution can be attempted at this stage.

To minimize any other inaccuracies due to a possible zero shift during and after installation, the following procedure was used: on May 14, 1973, water levels upstream and downstream of the cut-off, as measured by standpipe piezometers (Fig. 17), were nearly equal. Readings of all instruments at that date were considered as initial values and differences calculated for any subsequent change in water level. These differences were then added algebraically to the water levels recorded in standpipes on the date of the initial reading.

The pressure distribution observed at the middle of December 1973 after maintaining for one month steady state conditions on both sides of the cut-off is as follows:

a)   Within the accuracy of the instruments, the water pressure on the upstream face of the upstream wall increases linearly with depth. In other words, all instruments, except those at elevation 350'

which are in contact with core material, read the same elevation
as a standpipe piezometer (Piezometer No. 9) near the wall.

b)   The same pattern as for the upstream face of the upstream wall is
observed on the downstream face of the downstream wall.

c)   The pattern recorded on the upstream face of the downstream wall
is much more intricate. Readings for all piezometers are given on
Fig. 13 with contours at 5 foot intervals in the central section
which covers nearly all of the piles. Transverse walls separate
this section from the panel zones on both banks which must be considered separately.

The central section can be divided in two broad zones: above and
below about elevation 125'. Above this elevation there is still a
large hydraulic head whereas below this elevation the same water
level as on the downstream side of the downstream wall is recorded.

Variations in piezometric level in the upper part of the wall should
not be given too much significance since standpipe piezometer No.
202, which is located at elevation 280' between the walls, reads
about 7 feet more than the electrical ones in the same area. This
may be due to the inaccuracy of the electrical instruments or to the
time lag required for steady state conditions to be established in a
zone where a considerable amount of bentonite mud has been injected during construction of the cut-off.

Nothing much can be said about piezometric levels in the panel

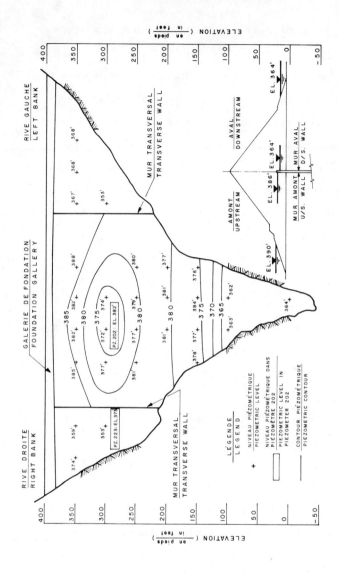

FIG. 13   NIVEAUX PIÉZOMÉTRIQUES SUR LA FACE AMONT DU MUR AVAL
PIEZOMETRIC LEVELS ON THE U/S. FACE OF THE D/S. WALL

zones, except that they are definitely lower than in the upper pile zone.

The pattern of piezometric levels described above indicates that the upper section of the upstream wall is relatively more pervious than the same section of the downstream wall and that the opposite occurs at the bottom of the canyon. In the latter case, the existence of one or more windows or openings at the contact of piles with bedrock could explain the difference between the overall permeability of the two walls. Such a possibility cannot be ruled out in view of the difficulties involved in the determination of bedrock elevation for each pile due to the presence of boulders or detached rock and the steepness of the abutments.

In the upper section, such an explanation is not valid and differences must exist between the watertightness of the walls themselves. However, the observed pattern could apply equally well to a watertight or leaky cut-off and other means must be used to estimate the cut-off efficiency.

Cut-off Efficiency

According to Marsal (1971), the efficiency of a cut-off can be defined in two different ways:

a)   As the ratio between the head loss through the cut-off itself and the overall hydraulic head across the dam.

b)   As the ratio between the reduction in seepage due to the cut-off and the seepage which would occur for the same head without a cut-off.

Since piezometric levels are more readily measured than seepage quantities, the tendency is to use the first definition. However, in the case of Manic 3, it is possible to make an assessment of efficiency on the basis of seepage reduction and this will be done before turning to the first definition.

Seepage Estimate with the Cut-off

A deep well system was installed at both cofferdams as shown on Figs 14, 15 and 16 to dewater the foundations for the dam embankment construction. This system had excess capacity for the actual conditions encountered and, during 1973 when the cut-off construction was completed, it was possible to vary the hydraulic gradient across the cut-off by changes in the rate of pumping. Variations in the discharge of the well systems at the upstream and downstream cofferdams as well as in water levels outside these cofferdams and in the foundation near the cut-off are shown on Fig. 17 for 1973.

For the first part of the year, interpretation of the results is diffi-cult because too many factors are varied at the same time. However, during August and September, a more significant test was carried out. This section of Fig. 17 is enlarged on Fig. 18 where four separate stages have been identified:

a)   The discharge upstream and downstream is adjusted to obtain

        steady seepage conditions.

b)   The steady state is maintained during 5 days.

c)   The downstream discharge is reduced to zero while the upstream

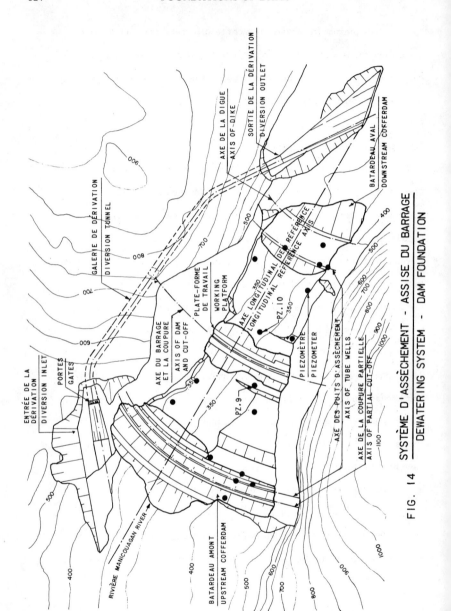

FIG. 14   SYSTÈME D'ASSÈCHEMENT - ASSISE DU BARRAGE
           DEWATERING SYSTEM - DAM FOUNDATION

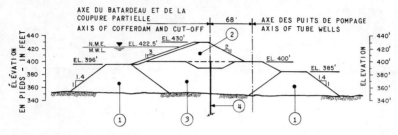

COUPE TYPE - TYPICAL SECTION

① DIGUE EN ENROCHEMENT
   ROCKFILL DIKE

② TILL

③ MATÉRIAU GRANULAIRE
   GRANULAR MATERIAL

④ COUPURE PARTIELLE
   PARTIAL CUT-OFF

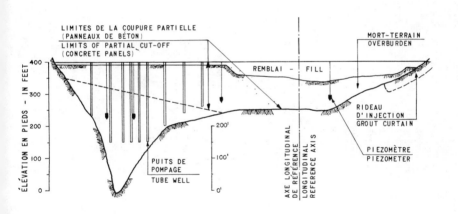

COUPE DANS L'AXE DES PUITS DE POMPAGE
SECTION ALONG THE AXIS OF THE TUBE WELLS

FIG. 15   SYSTÈME D'ASSÈCHEMENT - BATARDEAU AMONT
             DEWATERING SYSTEM - UPSTREAM COFFERDAM

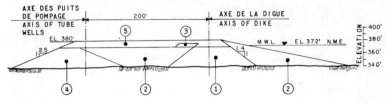

COUPE TYPE - TYPICAL SECTION

1. DIGUE EN ENROCHEMENT
   ROCKFILL DIKE

2. TILL DÉVERSÉ
   DUMPED TILL

3. TILL COMPACTÉ
   COMPACTED TILL

4. GRANULAIRE DÉVERSÉ
   DUMPED GRANULAR MATERIAL

5. GRANULAIRE COMPACTÉ
   COMPACTED GRANULAR MATERIAL

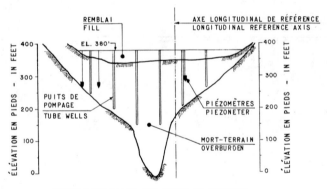

COUPE DANS L'AXE DES PUITS DE POMPAGE
SECTION ALONG THE AXIS OF THE TUBE WELLS

FIG. 16   SYSTÈME D'ASSÈCHEMENT - BATARDEAU AVAL
DEWATERING SYSTEM - DOWNSTREAM COFFERDAM

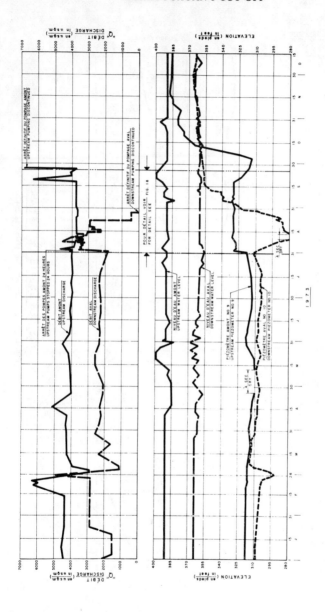

FIG. 17 RÉSULTATS DES ESSAIS DE POMPAGE
RESULTS OF PUMPING TESTS

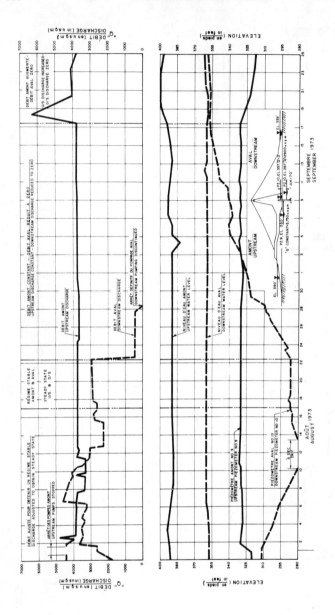

FIG. 18   ANALYSE DES ESSAIS DE POMPAGE
           ANALYSIS OF PUMPING TESTS

discharge is kept constant.

d) The upstream discharge is increased while the downstream dis-

charge remains zero.

The significance of this test lies in the third stage during which

the piezometric level downstream of the cut-off is allowed to rize by 70

feet while the discharge of the upstream pumping system is kept constant.

No measurable change in the water level upstream of the cut-off (upstream

piezometer No. 9) occurred during the whole of this period which lasted

26 days. Single measurements for this piezometer varied between 328'

and 332' with most values between 329' and 331'. The initial value was

330'. These small variations seem to have been brought about by fluctu-

ations in the upstream water level.

The fact that a change of 70 feet in the water level downstream of

the cut-off had no noticeable effect on the water level upstream of the cut-

off where the pump discharge was kept constant indicates that seepage

through the walls must be small. Assuming linearity and a possible error

of one foot in the water level measurement, as discussed above, the maxi-

mum seepage through the walls would not exceed the discharge rate at the

upstream cofferdam divided by the hydraulic head across that cofferdam or

3,750 usgm/60 $\approx$ 60 usgm. This rate of seepage applies for a hydraulic

head of 70 feet across the cut-off and, since the total head across the dam

will be 315 feet, a value of 60 x 315/70 usgm or 0.6 cfs is obtained. This

value represents a rough estimate of the upper limit of seepage through the

cut-off.

Seepage Estimate Without the Cut-off

   The dewatering data can also be used to estimate the amount of
seepage for a dam without cut-off. For this purpose, the readings of August 20, 1973, when a steady state had been established, will be used.

   At the upstream cofferdam, the hydraulic head is 59.5 feet for a
discharge of 3,750 usgm. Assuming a seepage path of 400 feet, the hydraulic gradient is $59.5/400 = 0.149$. For the dam, the hydraulic head
will be 315 feet for an approximate seepage path of 1,200 feet. This
yields a gradient of $315/1,200 = 0.263$ and the seepage Q through the dam
is obtained by simple proportion:

$$Q_{u/s} = \frac{3,750 \times 0.263}{0.149} = 6,610 \text{ usgm} = 14.7 \text{ cfs}$$

   This same calculation can be performed for the downstream cofferdam values. The hydraulic gradient is equal to $73.5/550 = 0.134$ for a
discharge of 2,900 usgm. Then, the seepage through the dam is:

$$Q_{d/s} = \frac{2,900 \times 0.263}{0.134} = 5,690 \text{ usgm} = 12.7 \text{ cfs}$$

Efficiency

   The two figures calculated above for the seepage are quite close
and this indicates that the foundation is fairly homogeneous throughout the
dam site. A rounded value (Q_0) of 14 cfs can then be used to estimate the
cut-off efficiency (E_Q):

$$E_Q = \left[\frac{Q_0 - Q}{Q_0} \right] 100$$

where Q is the seepage for a dam with cut-off
       Q_0 is the seepage for a dam without cut-off

$$EQ \geqslant \left[\frac{14 - 0.6}{14} \right] \ 100 \geqslant 96\%$$

The symbol equal to or larger than ($\geqslant$) is used in this equation since the estimated seepage through the cut-off represents an upper bound.

Another estimate of the cut-off efficiency can be made using piezometric data on Fig. 17. It has already been indicated that within the limits of their accuracy the electrical piezometers located on the outside faces of the walls read the same levels as the standpipe piezometers Nos 9 upstream and 10 downstream so that as a first approximation it can be assumed that the latter are located very near the walls.

After pumping was discontinued, the water levels gradually rose and on the downstream side of the cut-off reached tailwater level (elevation 364'). However, on the upstream side of the cut-off the piezometric level has stabilized at elevation 386', or nearly 4 feet below the upstream water level. Then, the piezometric efficiency (E_p) of the cut-off is approximately $E_p = \left[(26 - 4)/26 \right] \ 100 = 85\%$. If an additional foot of head loss is added to cover the distance between piezometer No. 9 and the cut-off (Fig. 19) then the efficiency drops to 81%.

This is a less encouraging result but the data used to arrive at this figure is not consistent. Since the pumping tests indicate that the foundation material is relatively homogeneous throughout the dam site, the hydraulic gradient downstream of the cut-off should be about the same as that upstream of the cut-off, which it is not. In January 1974, upstream

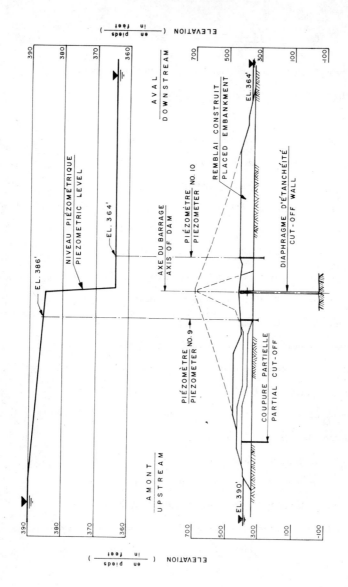

FIG. 19   NIVEAUX PIÉZOMÉTRIQUE LE 15 DÉCEMBRE 1973
PIEZOMETRIC LEVELS ON DECEMBER 15, 1973

piezometer No. 9 froze and after unfreezing and surging it was found that the water level at that point rose approximately to the upstream water level of 390 feet. Hence, the efficiency is definitely much higher than that calculated above.

Due to the small variations in water levels both upstream and downstream of the dam site, the efficiency cannot be calculated accurately at this stage but it certainly exceeds 90% which would correspond to a head loss through the alluviums of 2.5 feet.

Comparison with Other Cut-offs

The previous estimates of efficiency raise the question of what constitutes an acceptable efficiency or, in other words, what can be considered as satisfactory performance. The possibility of defective joints or windows at bedrock contact cannot be excluded and this may reduce the efficiency considerably. Ambraseys (1963) has shown that, for openings of not more than 0.1% of the cut-off area, a decrease in efficiency of 50% would occur.

Looking at this problem in another way, Ambraseys (1963 a) has shown that a cut-off which is 100 times less pervious than the surrounding material would have an efficiency of about 50%. This figure would rise to about 90% for an average permeability which is 1,000 times smaller than that of the surrounding material.

Case records for which the efficiency is calculated are not numerous, but the figures quoted are always high. Marsal (1971) reports an efficiency of 96% or more and negligible seepage for the concrete panel

cut-off at the José M. Morelos Dam. The reduction in seepage achieved
at the powerhouse cofferdam for the same project indicates an efficiency
of 98%. Finally, Dagenais and Turenne (1964) report an efficiency of 97%
for the concrete interlocking pile cut-off at the upstream cofferdam of the
Manicouagan 5 Hydro Power Development.

These few examples indicate that a 95% efficiency could be expected from this type of cut-off and that would correspond to a reduction
in permeability somewhat in excess of 1,000 times. The efficiency calculated on the basis of the estimated seepage is thus quite satisfactory and
it is expected that a similar efficiency will be obtained on the basis of
piezometric levels when data of sufficient accuracy becomes available for
such a calculation.

Extensometer Readings

A large amount of data has recently been obtained from both types
of extensometers installed in the cut-off walls. However, at this early
stage in the construction of the embankment, only the results of the four
benchmarks with dial extensometer will be reported as an indication of
the strains to be expected. These results agree with those of the electrical instruments and with the periodic surveys of the floor of the foundation
gallery which were started in August 1973.

The benchmarks with dial extensometer consist of a string of 1 inch
steel rods grouted in bedrock and extending to the foundation gallery where
they are topped with an oil filled dial extensometer. This instrument, which
is described by Casagrande and Poulos (1968), measures the total com-

pression of a wall at a given location. The settlement curve for each
instrument is shown on Fig. 20 together with the fill elevation on both
sides of the cut-off. A profile of the settlement along the axis of the cut-
off as of the beginning of this year is shown on Fig. 21. Three of the
instruments are located on the upstream wall, one in the center of the
canyon (pile 96) and two (piles 22 and 170) near the break in bedrock
profile on both abutments. The fourth one is installed in pile 296 of the
downstream wall, just in front of pile 96.

The settlement of the walls increased at a relatively modest rate
until a level corresponding approximately to that of the working platform
for the construction of the walls was reached. Afterwards, the settlement
increased at a much higher rate untill placement of fill was discontinued
for the winter. The average strain at each location for the measured
settlement as of February 4, 1974, is given in Table III.

TABLE III

Extensometer in pile No.	Length of pile (feet)	Settlement (inches)	Average strain (%)
22	171.5	0.215	0.0104
170	158.0	0.167	0.0088
96	425.0	0.561	0.0110
296	423.5	0.483	0.0095

These strains are almost equal for all piles even though their lengths
differ appreciably. This may be fortuitous since most of the load is due
to side friction on the walls which could be expected to increase with
depth. However, there are indications from the electrical extensometer

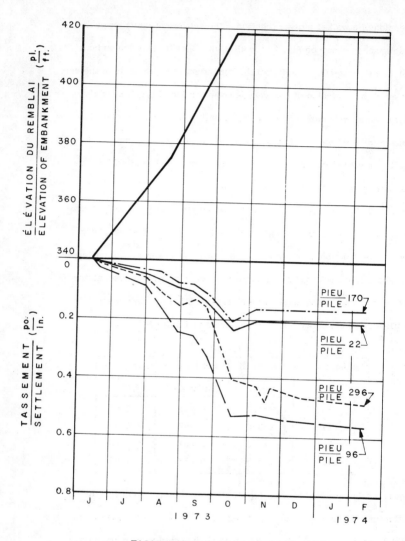

FIG. 20 TASSEMENT DU DIAPHRAGME MESURÉ PAR LES REPÈRES DE DÉFORMATION
SETTLEMENT OF THE CUT-OFF MEASURED BY BENCHMARKS WITH DIAL EXTENSOMETERS

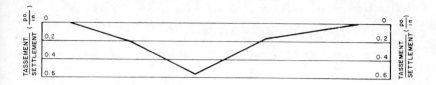

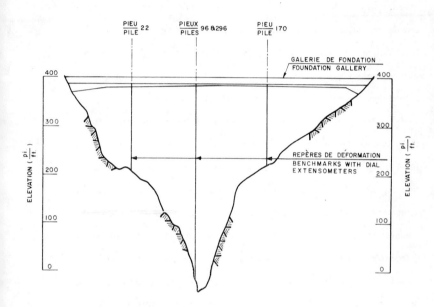

FIG. 21   PROFIL DU TASSEMENT DU DIAPHRAGME
SETTLEMENT PROFILE OF THE CUT-OFF

data that arching may occur in the lower part of the canyon and this could explain at least partially the results obtained.

If an average strain value of 0.01% is used and assuming a modulus of elasticity of 4.5×10^6 psi, an average stress of 450 psi is obtained. If the friction is assumed to increase linearly with depth, the stress would vary from 0 to 900 psi disregarding the small load due to the weight of the foundation gallery and the small height of fill on top of it. For a wall 2 feet thick and 160 feet deep this corresponds to an average side friction of 0.8 t/sq ft which is much more than the peak value of 0.24 t/sq ft recorded during a pulling test on a single pile 100 feet long.

Although the use of average values is always subject to caution, the previous analysis does indicate quite high stresses for an embankment level which is only 40 feet above the construction platform elevation. Unless the side friction reaches a peak value after a certain settlement, the stress at the bottom of the walls can be expected to exceed the nominal concrete strengths of 5,000 psi when the additional 270 feet of fill required to complete the dam are placed. Thus numerous diagonal shear planes with small displacement may form in the walls, but this would not produce any open cracks and hence should not affect significantly the efficiency of the walls.

Other Instruments

Little can be said about other instruments installed in the cut-off walls. Data from inclinometer casing surveys are not yet available and shear strips have not indicated any failures.

The installation of instruments around the foundation gallery is complete, but not enough fill has been placed to obtain meaningful results.

CONCLUSION

At Manicouagan 3, the designers were faced with a foundation problem which is becoming more and more frequent as the more favourable sites for hydro power are being developed. Seepage through the 400 feet of sediments beneath the embankment is controlled by means of a positive cut-off consisting of twin walls of interlocking concrete piles and panels extending to bedrock and topped with a gallery. Advantage was taken of the need for a partial cut-off at the upstream cofferdam to provide additional seepage control by linking this second cut-off with the core of the dam by means of horizontal blanket. A horizontal drainage layer is incorporated in the downstream shell to collect seepage if the main cut-off became seriously defective.

Cut-offs of interlocking piles or panels have been built at numerous locations but always as single walls and to depths not exceeding 250 feet. To reach a depth of more than 400 feet and maintain the continuity of the walls new tools were designed. These made possible frequent control of the verticality of the piles and corrections in alignment where required. Inspection of the top 40 feet of the walls after excavation of the working platform showed continuous walls with narrow bentonite filled joints, except at two locations where repairs were carried out.

Considerable data has already been accumulated on the performance

of the cut-off which is thoroughly instrumented. Although the small hydraulic head across the cut-off does not allow very accurate estimates of
efficiency, the low rate of seepage (less than 0.6 cfs) indicated by preliminary analyses and the small or zero gradient recently measured on
both sides of the cut-off give confidence that performance will be satisfactory when the reservoir is filled at the end of 1975.

Measurements of deformation of the walls already indicate fairly
high compressive stresses (maximum 900 psi) for an embankment level
which is only about 40 feet above the platform from which the walls were
constructed. An estimate of the maximum stress to be expected should be
possible after the addition of about 140 feet of fill on top of the cut-off
this coming season.

ACKNOWLEDGEMENTS

The Manicouagan 3 Hydro Power Development is designed
by Asselin, Benoît, Boucher, Ducharme, Lapointe Inc., Consultants, in cooperation with the engineering staff of Hydro-Quebec.
The Board of Consultants for the main dam consists of Professor
A. Casagrande, Chairman, Dr. F.A. Nickell, Mr. F.B. Slichter and
the late Mr. R. Peterson. Mr. S.D. Wilson is consultant on instrumentation. The Board members took an active part in the design.
Many of the important features of the main dam and in particular
the novel ones, were contributed by them.

Thanks are due to the owner, the Quebec Hydroelectric
Commission, for permission to present this paper.

REFERENCES

- Ambraseys, N.N., (1963) : Cut-off Efficiency of Grout Curtains and Slurry Trenches - Symposium on Grouts and Drilling Muds in Engineering Practice, Butterworths, London.

- Ambraseys, N.N., (1963 a) : Author's Reply on Cut-off Efficiency of Grout Curtains and Slurry Trenches - Symposium on Grouts and Drilling Muds in Engineering Practice, Butterworths, London.

- Benoît, M., Crépeau, P.M. et Larocque, G.S., (1967) : Influence des fondations sur la conception du barrage Manicouagan 3 - Neuvième Congrès des Grands Barrages, Istamboul - Volume I - Q.32, R.48.

- Casagrande, A., (1969) : Earth and Rockfill Dams - Main Session 3, Discussion on Seepage Control - Proceedings of the Seventh International Conference on Soil Mechanics and Foundation Engineering, Mexico - Volume III - pp. 278-283.

- Casagrande, A., and Poulos, S.J., (1968) : Design and Installation of Benchmark Equipped with Dial Extensometer - Pierce Hall, Harvard University, Cambridge, Massachusetts.

- Dascal, O., (1973) : Manicouagan 3, Coupure étanche de la fondation. L'Ingénieur, Novembre 1973, N° 295.

- Dagenais, C.A. et Turenne, G., (1964) : Analyse et solution du problème d'étanchéité au batardeau amont de l'aménagement de Manicouagan 5. The Engineering Journal, January 1964.

- Dréville, F., Paré, J.J., Capelle, J.F., Dascal, O., Larocque, G.S., (1970) : Diaphragme en béton moulé pour l'étanchéité des fondations du

barrage Manicouagan 3 - Dixième Congrès des Grands Barrages,
Montréal - Volume II - Q.37, R.34.

- Marsal, R.J. and Resendiz, D., (1971): Effectiveness of Cut-offs in
Earth Foundations and Abutments of Dams - Proceedings of the Fourth
Panamerican Conference on Soil Mechanics and Foundation Engineering
- Volume I - pp. 237-312.

- Paré, J.J. and Ciurli, S., (1971): The Manicouagan 3 Cut-off Walls -
Cancold Annual General Meeting, Vancouver.

- Pigeon, Y., (1972): Manic 3 Foundations - Paper Presented to the Can-
adian Electrical Association, Hydraulic Power Section, Spring Meeting,
Montreal.

UNUSUAL FOUNDATION DEVELOPMENTS
AND CORRECTIVE ACTION TAKEN
By
Doyle L. Christensen, Regional Engineer
Division of Safety of Dams
Department of Water Resources
The Resources Agency
State of California

At two minutes to midnight on March 12, 1928, the
St. Francis Dam, 37 miles northwest of Los Angeles, failed. It
released 38,000 acre-feet of storage. The dam caretakers, the
powerhouse operators, and about 450 other people downstream of
the dam lost their lives as the water rushed 52 miles down the
Santa Clara River to the Pacific Ocean. Property damage amounted
to several million dollars.

The St. Francis Dam was a 205-foot-high concrete gravity
dam. The Commission of engineers and geologists that investigated
the failure and reported to the Governor found "The failure of
St. Francis Dam was due to defective foundations". The right
abutment was a reddish conglomerate which the Commission reported
quickly softened and changed into a mushy mass when immersed in
water. They concluded that the foundation apparently softened as
the reservoir filled, and either a blowout or settlement, or both,
precipitated the failure.

As the result of the St. Francis Dam failure a law was
passed in 1929 providing State supervision of dams for safety in
California. This law is now enforced by the Division of Safety
of Dams in the Department of Water Resources. All dams that are

343

25 feet or more in height or store 50 acre-feet or more of water
are subject to State jurisdiction. Federal dams are excluded.
Currently the Division has over 1,100 dams under its jurisdiction.
Our office makes an independent review of all plans and specifications of dams to be constructed, repaired, altered or removed.
After we are satisfied that the plans and specifications are satisfactory with respect to safety, they are approved. The work in
the field is then supervised by our Field Engineering Branch to
see that the approved plans are followed. I supervise the
Division's field activities in Northern California.

From the inception of the law we have been very critical
of dam foundations. It is our standing practice to inspect and
approve all foundations of a dam and its appurtenances before
they are covered with concrete, soil, rock, or whatever. Over
the years we have had some out-of-the-ordinary experiences related
to foundations, and this morning I would like to describe three
of them to you. I have several slides I would like to show along
with the discussion, so could we now have the lights turned off
please.

(Slide 1) The first case concerns a concrete arch dam
that is 635 feet high and stores just under one million acre-feet
of water. Construction was started in 1966 and finished in July
1969. (Slide 2) Bedrock at the site is a series of meta-volcanic
rocks known as the Oregon City formation. (Slide 3) The rock is
greenish gray, hard, dense, massive, and moderately jointed.
(Slide 4) These slides show the general nature of the rock. The
excavation was carried well into fresh, ringing bedrock as shown

Slide 1

Slide 2

Slide 3

Slide 4

by this excavation for the arch seat. (Slide 5) The joints were
generally tight. The grout curtain depth varied from 100 feet at
the ends of the dam (Slide 6) to 300 feet in the streambed area.
Considerable consolidation grouting was done but the take was light.
(Slide 7) The number of holes in each block varied, depending on
bedrock conditions but in some areas were as close as 10-foot center
both ways. (Slide 8) Overall about 118,000 lineal feet of grout
holes were drilled with a take of about 27,000 sacks, or less than
1/4 sack per lineal foot of hole. About 200 foundation drain holes
were drilled downstream of the grout curtain carried to a depth of
about 75 percent of the grout curtain.

        (Slide 9) After completion of the dam in 1969, we started
making our periodic inspections, inside and outside, of the dam as
the reservoir filled to capacity. Leakage from the foundation
drains totaled about 1.5 cfs with the reservoir full. The unexpected development came in 1972, just three years after the dam
was finished. It was discovered in connection with an investigation
of sand coming from one of the drain holes. In sounding the hole
to check on the sand, it was found that the hole was open only for
a depth of 62 feet. This compared to an original depth of 260 feet.
In trying to flush out the sand the obstruction was found to be
solid. Sounding the 200 drain holes, it was found that 36 were
plugged with calcite. (Slide 10) A little surprisingly, the plugged
holes were all in the canyon bottom except for two or three
scattered up each abutment. These views show the streambed area.
(Slide 11) In general, the plugged zone in the holes was 50 feet
below the gallery floor, or about 20 feet into bedrock. The

Slide 5

Slide 6

Slide 7

Slide 8

Slide 9

Slide 10

plugged zone usually varied in length from 2 to 5 feet. (Slide 12)
Some of the calcite was cored. These two slides show the type of
material cored. (Slide 13) I will also pass this sample around
for you to inspect. The material is typical of that found in
galleries of concrete dams which has leached out of the concrete
and deposited at joints and cracks. Apparently the drain holes
were plugged by the leaching of the grout from the grout curtain
and the consolidation grouting. A drill crew was called in to
ream out the holes to their original depth and wash them thoroughly.
Three men did the job in two weeks at a cost of about $6,000.
(Slide 14) The drains will be checked from time to time in the
years to come to be sure they are open and functioning properly.
At the same time, it has alerted us to watch for similar problems
at other dams in the State. We have found lots of drain holes
plugged to varying degrees at other dams in the State over the
years, but seldom this soon after construction. The sand in the
drain, incidentally, we concluded was washed into the hole from a
gallery cleanup operation. We were concerned in the beginning that
it might be coming from the reservoir through an opening in the
foundation.

          Along this same line, we have found "popcorn" or porous
concrete used in connection with drainage systems to be similarly
troublesome. (Slide 15) The thrust block at this dam had an
extensive drainage system including a "popcorn" concrete drain
to carry leakage from a prominent joint. Calcite leached heavily
from the popcorn drain depositing material 3 inches thick in the
bottom of the outfall pipes, and after nine years plugged several
vertical drain holes in each outfall pipe. Note also the calcite

Slide 11

Slide 12

Slide 13

Slide 14

Slide 15

Slide 16

below the outfall drains. The vertical drain holes had to be
reamed out. It was difficult to get a tool into the vertical
holes, but it was finally accomplished using a pipe for a guide
and a plumber's Roto-Rooter.

The second dam that developed a foundation problem was
at this site. (Slide 16) The view is downstream. Dam crest elevation is at the top of the clearing. The dam was built over a two-
year period in 1958 and 1959. The design called for a dam 95 feet
high with a storage of 1,800 acre-feet for a city's domestic water
supply. (Slide 17) The fill material was a rocky clay residual
soil from the reservoir hillsides. The finer grained material was
used in the core and the rockier material was routed to the outer
slopes. Drainage features included a blanket drain and filter in
the streambed, and a 3' x 3' chimney drain up each abutment. (Slide 18)
Bedrock was a volcanic formation consisting mainly of tuff in the
core area and some flow rock under the shells. An attempt was made
to grout the foundation, but it was abandoned because of heavy
leakage around the nipples under very light pressure. Deepening
of the cutoff trench to tight bedrock was done in lieu of grouting.
(Slide 19) A typical view of the blocky tuff is shown here. (Slide 20)
Cleanup in the core was carefully done by hand mainly using air.
Usually a few feet of foundation was prepared at a time because the
tuff air slacked quite badly. (Slide 21) The core material at the
abutment contacts was placed on the wet side of optimum so it would
work well into the irregularities of the rock surface. Hand compaction and wheel rolling was done along the abutments. (Slide 22)
The lower left abutment looked like this (Slide 23), and the upper
left abutment looked like this.

Slide 17

Slide 18

Slide 19

Slide 20

Slide 21

Slide 22

Slide 23

Slide 24

The construction was stopped during the winter of 1958,
30 feet below top of dam, and resumed in the spring of 1959, and
completed by early summer. The reservoir filled for the first
time in February 1960, and in short order a seepage area developed
high on the downstream face adjacent to the left groin as shown
in this slide (Slide 24). The area is about at the level where
the fill was stopped in the winter of 1958 and extended out from
the groin about 70 feet. Drain trenches about 4 feet deep with
pipes and filter material were placed in the wet area and carried
over to and down the left groin. Leakage from this area amounted
to about 40 gpm with the reservoir at spillway level. Foundation
leakage from the rest of the dam was only 15 gpm. We in the
Division felt the leakage was passing through the tuff foundation
and finding its way through pervious shell material to the downstream face. This leakage condition was checked closely during
our annual inspections, but no piping, sloughing, or other problems
ever developed. In 1970, the owner advanced a proposal to raise
the storage level 7 feet to acquire more water. We said the left
abutment leakage problem would have to be corrected if the enlargement were to be made. (Slide 25) The owner's engineer working on
the enlargement plans put several exploratory holes, including
30-inch bucket auger holes, in the left end of the dam to investigate
the leakage condition. The bucket auger holes were entered and inspected but the observations were inconclusive. The engineer subsequently concluded that the leakage was coming mainly through the
embankment materials and proposed chemical grouting. We told the
engineer we felt the leakage was coming through the foundation

Slide 25

Slide 26

rock and that cement grouting seemed more appropriate. Eventually
we approved the engineer's plans with the understanding that the
foundation would be grouted if the embankment grouting proved
unsuccessful in solving the problem.

(Slide 26) Chemical grouting was started in May 1973.
These slides show the location of the holes and the equipment used.
Twenty-three holes were placed over a distance of 120 feet at the
left end of the dam. (Slide 27) The holes varied in depth from
30 to 84 feet. The spacing was started out on 16-foot centers and
split down to 4 feet in some places. The take varied in the holes
from 24 gallons to a maximum of 663 gallons. The total take was
3,241 gallons of grout. Pressures were kept low to avoid any
uplift and generally limited to about 5 psi at the nipple. (Slide 28)
If water losses developed as the holes were drilled, they were
grouted. AM-9 grout was used. Gel times varied from 5 minutes
to 15 minutes. Dye tests were used in the beginning with the water
tests, but dye never did show -- either in the leakage area, in
the reservoir, at the downstream toe, or anywhere else. Some grout
was placed in the embankment and possibly along the contact, but
most of it was placed in the foundation tuff. The results were
quite successful. The leakage from the left groin was 37.5 gpm
at the beginning of the grouting. The grouting was completed in
30 days and by that time the leakage had dropped to 3.7 gpm. A
little of this was attributed to a 2-foot drop in the reservoir
level during the grouting. Also the seepage area dried up almost
completely on the surface of the fill. The piezometers and wells
placed to monitor the results were gradually showing a drop in

Slide 27

Slide 28

the piezometric level and have continued to do so. We have concluded that it is now safe to make the enlargement; however, the
piezometric levels within the embankment will be closely monitored.
The grout work cost about $24,000.

This experience was interesting to me for two reasons.
First, the chemical grouting proved to be very effective in grouting
the tuff bedrock, the rocky clayey embankment materials, and
possibly the fill-abutment contact. Secondly, it demonstrated to
me again that our general policy of not directing the approach to
problems, but rather looking for safe end results, is a proper
policy. On occasion we may not agree with an engineer's design
or solution to a problem, but we try to avoid imposing our preferred
solution on the engineer and approve his plan if the safety of the
dam is not adversely affected.

(Slide 29) The third dam and foundation problem I want
to discuss has to do with another domestic water supply for a
small town near the Oregon line. The height of the dam is 35 feet.
The reservoir stores 535 acre-feet. The dam is a central core
earthfill with transition zones upstream and downstream and (Slide 30)
shells of coarse dredger tailings. (Slide 31) The site is along a
creek where gold sluicing and dredging was done many years ago.
(Slide 32) A core trench 45 feet deep was required to get below
the tailings and other pervious material. (Slide 33) These slides
show the core trench at various stages of the excavation. (Slide 34)
Special care was taken in cleanup (Slide 35) of the core trench and
the placement of backfill to assure an impervious foundation barrier.
The completed dam looked like this (Slide 36).

Slide 29

Slide 30

Slide 31

Slide 32

Slide 33

Slide 34

Slide 35

Slide 36

On the first filling of the reservoir we watched the
dam very closely as is our normal procedure. When the reservoir
was full, it was performing normally in all respects with leakage
of 30 gallons per minute downstream of the dam. After being
full about 2 months, we received a call from the owner's engineer
that the reservoir level was dropping one foot per day or losing
water at the rate of 12 cfs for no apparent reason. The dam was
immediately checked. There were no apparent problems right at
the dam, but a flow of over 100 gallons per minute had developed
in the creekbed 450 feet downstream of the dam, and other wet
areas downstream had developed. Also bubbling was occurring in
the reservoir 150 feet upstream of the right end of the dam. In
addition, an old gold mine shaft one-half mile below the dam was
rapidly filling up with water. The shaft was used by the City for
part of their water supply. It soon became apparent that the
reservoir was probably losing water to old mine drifts in the area.
The outlet gate was opened to drain the reservoir and when the
water level was 9 feet below spillway level this hole (Slide 37)
in the reservoir floor was visible. The dam embankment is to the
right of the picture. A second hole 30 feet upstream was also
found. The first hole was the only one leaking water. It was
finally concluded that the two old mine shafts, possibly air
shafts, had over the years been filled with loose material and
forgotten about. Even the borrow excavation in the area for
tailings and the travel of the heavy equipment gave no clue to
the existence of the shafts. The filling of the reservoir and
the seepage of water down the first shaft was enough to break

the plug. We checked the depth of the first hole and it was
44 feet to the top of fill material with the shaft continuing
deeper. After the reservoir had been pulled down below both holes,
the second hole was excavated down 22 feet and backfilled with
impervious material. The first hole was filled to within 9 feet
of the surface with impervious core material, then a 5-foot-thick
tapered concrete plug was poured followed by 4 feet of compacted
impervious material. The reservoir was raised again and there has
been no problem with leakage since then -- 1960 -- to the present
date.

     The only lesson to be learned here, I suppose, is that
even though considerable engineering care may be taken to overcome
some difficult site problems, you might still be surprised later
by something like an old mine shaft.

Slide 37

EMERGENCY GROUTING OF OLD RIVER

LOW SILL STRUCTURE, LA.

by E. Burton Kemp*

Old River Low Sill Structure, (Fig. 1), located approximately 154
miles upstream of New Orleans, La., on the westbank of the Mississippi
River, was completed in 1960 as part of a project to prevent the capture
of the Mississippi River by the Atchafalaya River.

A study, conducted by the U. S. Corps of Engineers in 1952, established that the Atchafalaya River would capture over 40 percent of the
flow of the Mississippi River by 1971 if steps were not taken to prevent
this. Once this capture occurred, it would only be a matter of time before
the Atchafalaya River would be carrying the majority of the Mississippi
River flow with disastrous results down river on both the Mississippi and
Atchafalaya Rivers.

Principal features of the plan consist of an overbank structure to
accommodate high water flows; a Low Sill Structure to control the percentage of flow between the Mississippi River and the Atchafalaya River;
a lock for navigation between the Mississippi, Red, and Atchafalaya Rivers;
and a closure dam to seal off Old River. The entire project was completed
in July 1963.

(Fig. 2). During the 1973 flood, an eddy developed behind the south-
east wing wall and on April 12, 1973, the major portion of this wing wall
collapsed. (Figs. 3 & 4). Surveys in the forebay area indicated that a
large scour hole had developed in front the gate bays 8, 9, 10, and 11 to
a maximum depth of 65 feet. Immediate remedial action consisted of the

---

*District Geologist, Chief, Geology Section, Corps of Engineers,
New Orleans, La.

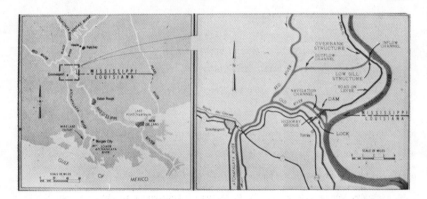

Figure 1

Figure 2

Figure 3

Figure 4

construction of a rock dike at the south end of the structure to temporarily replace the collapsed wing wall, and placing approximately 100,000
tons of riprap in the forebay area in the scour hole. (Fig. 5).

Since the sheet pile beneath the structure in the forebay area extended
to an elevation of -36 feet MSL, and since the scour hole had developed to
an elevation of -65 feet MSL (Fig. 6), it was determined that a comprehen-
sive investigation of the foundation beneath the structure and the stilling
basin was needed.

Under the direction and supervision of the New Orleans District, a detailed boring program, using Mobile District crews, and drill rigs mounted
on self-elevating drill barges (Fig. 7); on trucks (Fig. 8); on skid platforms (Fig. 9), and on spud barges (Fig. 10), was initiated at the Low Sill
Structure on 22 September 1973. Drilling operations were extremely complex
because of: operating on a 24-hour basis; providing for one lane of traffic
on Louisiana Highway 15 to be open to vehicular traffic at all times; and the
continuous high water with accompanying high velocities and drift material.
In addition, continuous coordination resulting from constant gate manipula-
tions was required to insure the continuity of the drilling and the maintenance of the stand pipes necessary for grouting. The gate operations were
the result of the concurrent operations consisting of: construction of a
large rock dike at the south end of the structure to replace the collapsed
wing wall; removal and leveling of rock in the forebay area; maintaining the
approximate 30 percent flow through the structure; prevention of excessive
head differentials; and finally, prevention of bank caving and bottom scour-
ing downstream from the structure. Since that time, a total of 48 borings

Figure 5

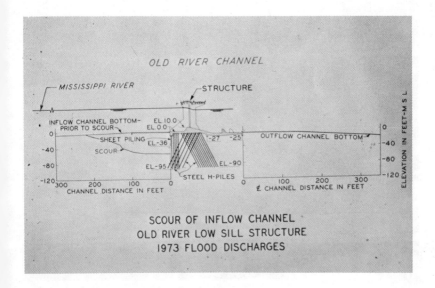

Figure 6

Figure 7

Figure 8

Figure 9

Figure 10

have been drilled. (Fig. 11). This includes borings along the B line
which is upstream of the centerline of the structure and downstream of the
upstream sheet pile; along C line located downstream of the structure and
upstream of the sheet pile between the gated structure and the stilling
basin; along D line located downstream of the centerline of the structure
and downstream of the sheet pile between the gated structure and the stilling basin, and along E and F lines located downstream of the centerline of
the structure and downstream of the sheet pile between the gated structure
and the stilling basin. These borings were drilled through the structure
and stilling basin concrete slab in gate bays 4, 6, 7, 8, 9, 10, and 11.
(Fig. 12.) In addition, one boring was drilled in the first non-gated bay
on the south side of gate bay 11. The borings indicated a cavity, as shown
on figure 12, beneath the structure and stilling basin that varied in size
from a maximum of 52.2 feet in boring 11 B to 1.8 feet in boring 7 D''.
Along the B line, (Fig. 13), the cavity varied in size from 6 feet in 7 B
to 52.2 feet in 11 B. Along the C line, (Fig. 14), the size of the cavity
varied from 1.9 feet in 6 C'' to 44.7 feet in 9 C'; and along the D line,
(Fig. 15), the size of the cavity ranged from 1.8 feet in 7 D'' to a maximum of 31.9 feet in 9 D. No cavity was detected along the E and F lines of
borings. Generally, the cavity consisted of an upper, open zone of water,
underlain by a very soft muck material. However, in several borings, no
muck was encountered. The muck material consisted generally of saturated,
loose, silty sand and silt, with varying amounts of clay. Beneath the
muck material in the B and C line of borings was firm foundation material
consisting of silty sand and clay, while foundation filter sand with traces

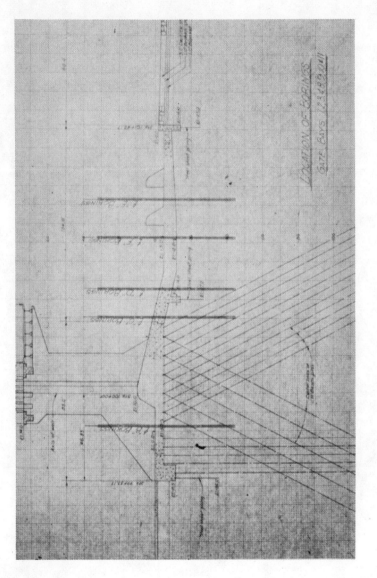

Figure 11

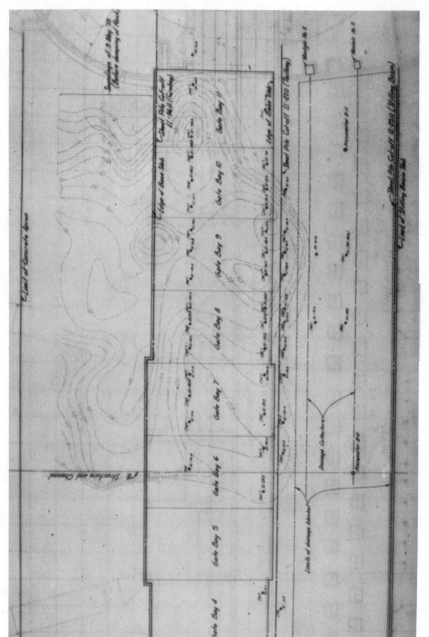

Figure 12

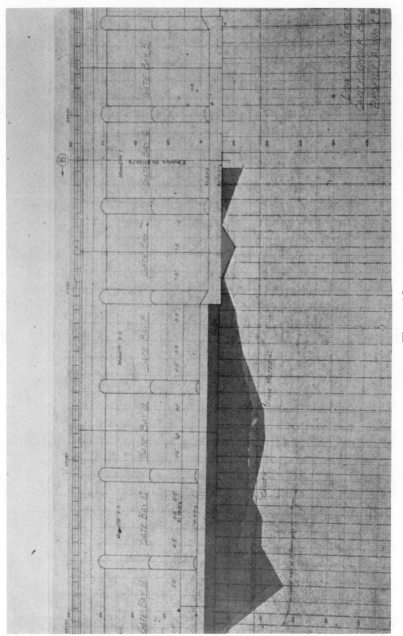

Figure 13

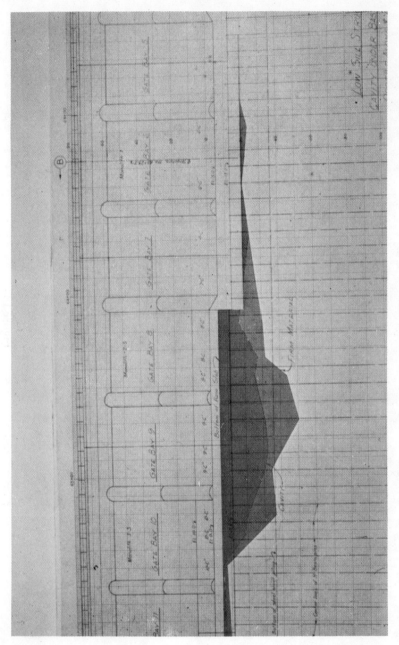

Figure 14

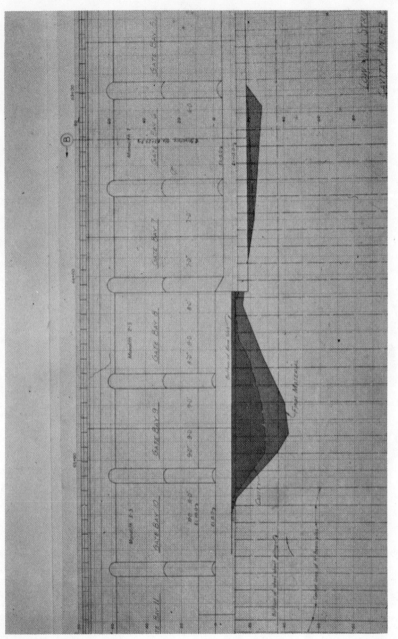

Figure 15

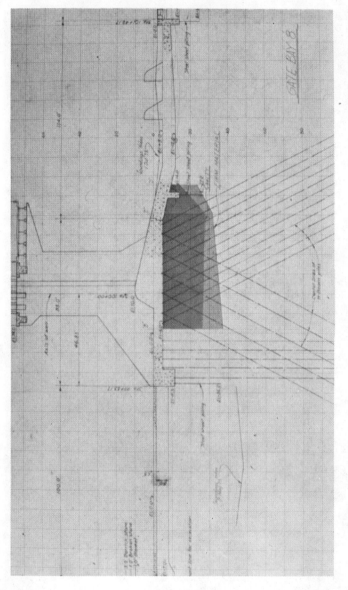

Figure 16

of pea gravel were encountered along the D line of borings.

Initial uplift pressure readings along the B and C line, (Fig. 11),
made as each boring was completed, indicated that the uplift pressure beneath the structure was equal to the headwater pressure. Initial pressure
readings in the D line of borings, (Fig. 11), indicated that the uplift
pressure behind the sheet pile and beneath the stilling basin was almost
equal to the headwater pressure. This indicated either that the sheet pile
between the structure and the stilling basin was ruptured or contained openings, or as indicated by the C and D borings, that the cavity extended beneath the base of the sheet pile which was at -27 feet MSL. (Fig. 16).

Since the boring program indicated a cavity of such size (estimated
30,000 CY), it was determined to initiate a grouting program immediately.
A grout mix was designed, in close cooperation with personnel from the
U. S. Army Waterways Experiment Station, with the following characteristics: self-leveling; a density of 115 lbs, sufficient to keep it in place;
a minimal amount of cement material to prevent hanging on the steel support
and batter piles beneath the structure; sufficient strength after set-up
to prevent erosion by water; and capable of providing lateral support to
steel piling. A contract was let, and on 23 November 1973, under the direction and supervision of the New Orleans District, grouting operations were
initiated at the Low Sill Structure, using squeeze-crete (Fig. 17) pump
units and blender pots, (Fig. 18), with mono and Gardner-Denver Duplex
pumps. As of this date, a total of just over 32,000 CY of grouting material
has been pumped into the B, C, and D line of borings at the Low Sill Structure, using these pumps. Three basic types of grout mixtures have been

Figure 17

Figure 18

Figure 19

Figure 20

used, coded as OR-5, OR-13, and OR-23. All contain varying amounts of
barite, bentonite, cement, and water. The OR-13 is a non-sanded mixture
used in the initial stages of grouting in an attempt to incorporate the
muck material into the grout and prevent the stripping out of sands contained in the other mixes. The OR-5 is a sanded mix, used after the initial pumping with OR-13. Both OR-5 and OR-13 contain only about 3 to 6
sacks of cement per cubic yard, while the OR-23 mix contains approximately
12 sacks of cement and is used as a topping-off mixture. The OR-23 mix
was designed to produce a good bond between the bottom of the slab and
the grout, and in addition, provide a hard capping to help prevent erosion
by channeling.

In order to adequately monitor uplift pressures and grout levels, a
sophisticated instrumentation program was developed and monitored by personnel of the Engineering Physics Branch, Concrete Lab, WES. The uplift
pressures were measured using CEC strain gage pressure gages with a range of
0-50 PSI, (Fig. 19), installed in waterproof housings, and located immediately beneath the concrete slab. Grout monitoring devices consisted of two
electrodes spaced approximately 1 - 1½ inches apart, protruding from an
encapsulated body containing the electrode wires, and also steel balls
necessary to provide weight (Fig. 19). These gages measured the different
conductivity exhibited by air, muck, water, and grout, and were calibrated
such that full scale (10-inch) deflection on the recorder, (Fig. 20), was
obtained with the age in air, and half scale (5-inch) deflection with the
gage inserted in water. With this calibration, grout detection was indicated
by a 1 inch deflection from zero, water contaminated with cement by a 1 - 1½

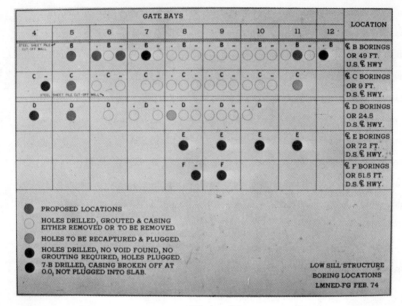

Figure 21

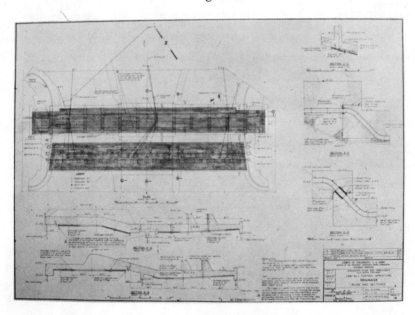

Figure 22

inch to 3 inch deflection, and muck by a 3 to 4 inch deflection. Strong
water current was indicated by a 6 - 7 inch deflection.

Prior to the placement of any grout material, water was pumped through
the grout pipe and discharged approximately 1 foot from the bottom of the
muck material in an attempt to cause as much muck as possible to be in
suspension. This water pumping action was followed immediately by the injection of OR-13, the unsanded mix, in an attempt to effect a better composite mix between the OR-13 and the existing muck material and create a
type of soil cement. Without this incorporation of the muck into the grout
mix of bentonite, barite, and cement, any sand injected into the muck at
this time might be stripped out, resulting in a mounding of sand material
in the immediate vicinity of the discharge pipe. Approximately 300 Cy of
OR-13 mix was pumped into B and C lines of borings, (Fig. 21), and was followed immediately by OR-5. Initial pumping rate was approximately 40 CY
per hour, and was increased gradually to a maximum of 60 yards per hour.
This slow rate of pumping was necessary to avoid a pressure buildup beneath
the gated structure that would exceed 10 PSI above the headwater pressure,
and beneath the stilling basin that would exceed 4 PSI above the tailwater
pressure.

When grout elevations reached the -20 foot level in B and C line of
borings, grout monitoring devices suspended in the D line of borings began
to indicate the presence of grouting material along the D line of borings
behind the sheet pile and beneath the stilling basin. (Fig. 21). Apparently,
this was the result of the surge of grouting material beneath the sheet pile.
This initial surge of grout material beneath the sheet pile resulted in the

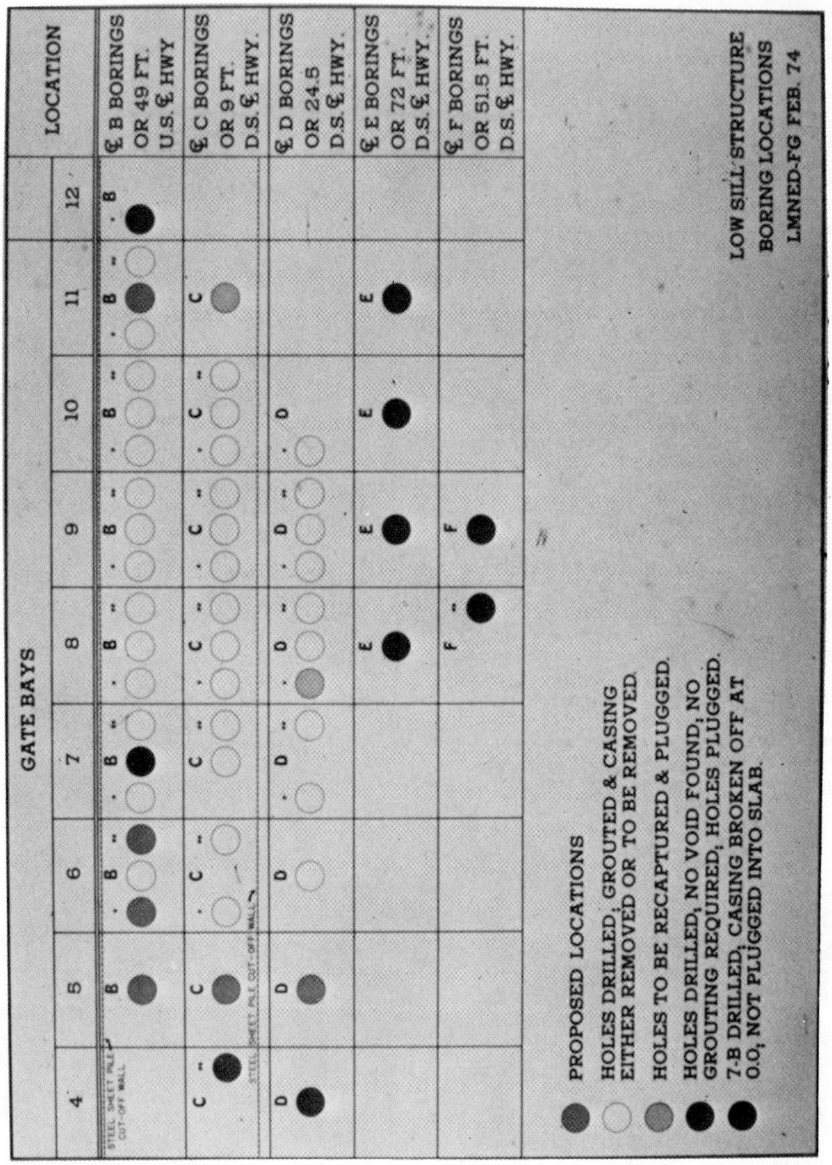
GATE BAYS
LOCATION
4
5
6
7
8
9
10
11
12
STEEL SHEET PILE CUT-OFF WALL
℄ B BORINGS OR 49 FT. U.S. ℄ HWY
℄ C BORINGS OR 9 FT. D.S. ℄ HWY.
℄ D BORINGS OR 24.5 D.S. ℄ HWY.
℄ E BORINGS OR 72 FT. D.S. ℄ HWY.
℄ F BORINGS OR 51.5 FT. D.S. ℄ HWY.
PROPOSED LOCATIONS
HOLES DRILLED, GROUTED & CASING EITHER REMOVED OR TO BE REMOVED.
HOLES TO BE RECAPTURED & PLUGGED.
HOLES DRILLED, NO VOID FOUND, NO GROUTING REQUIRED, HOLES PLUGGED.
7-B DRILLED, CASING BROKEN OFF AT 0.0, NOT PLUGGED INTO SLAB.
LOW SILL STRUCTURE
BORING LOCATIONS
LMNED-FG FEB. 74

Figure 21

placement of enough grout material behind the sheet pile to raise the level
of the grout in the D line of borings to an elevation of approximately -17
feet MSL. As this surging of grout material beneath the sheet pile and
into the D line cavity beneath the stilling basin was occurring, there was
a noticeable drop in the uplift pressure along the D line of borings. A
setup period was effected in order to allow sufficient time for the grout
mixture to set, and after 5 days, grouting resumed in the B and C line of
borings with OR-5 grout mix at a rate of about 60 yards per hour and continued until the level of grout in these lines was raised to an elevation
varying between -15 and -18 feet MSL. At this point, it was first noticed
that a buildup of grout was not occurring in the 8 C' boring area. (Fig. 21)
A careful examination of borings along the D line, and in particular, in the
boring 8 D' and 8 D, revealed that no additional buildup of grout had occurred in the D line of borings since the initial surge, indicating that the
base of the sheet pile was effectively anchored in grout and that no ruptures
or openings existed in the sheet pile behind gate bays 8, 9, and 10.

In an attempt to prevent loss of grout in 8 C', first SICA 4A and
then Reg-Set, both accelerators, were added to the OR-5 mix and pumped into
8 C'. However, both were unsuccessful in stopping the leakage in 8 C', and
in an unsuccessful attempt to determine where the leakage was occurring, a
large quantity of dye was injected into 8 C'. During this entire operation
of accelerated pumping into 8 C', continual monitoring of the grout level
and the pressure level along the D line borings, particularly in the 8 D'
and 8 D, was carried on. At no time was there any indication of an increase
in the uplift pressure or an increase in the level of the grout in the D

line of borings. In order to equalize pressures on both sides of the sheet pile between the C and D borings, grouting operations were initiated in the D line of borings behind gates 8, 9, and 10 (Fig. 21) using the OR-5 mix, and continued until the level of the grout had reached an elevation of -13 feet MSL, or one (1) foot below the base of the stilling basin slab. The uplift pressures beneath the stilling basin dropped to tailwater pressure and remained at this level for the duration of the grouting operations. Grouting operations were then continued into the C line of borings. In another attempt to seal off 8 C', approximately 500 CF of foam rubber squares, measuring 1' x 1' x 6" were forced down the grout pipe into 8 C'. Immediately behind this foam rubber was pumped a mixture of diesel fuel, bentonite, and cement, referred to as the DOC mix. In addition, several hundred pounds of IMCO flakes were also mixed in with the DOC and injected into hole 8 C'. Immediately following the injection of approximately 31 CY of DOC mix, grouting operations were resumed in 8 C' using OR-5 mix, with Reg-Set and IMCO flakes added. The level of grout in 8 C' began to rise, and when it reached an elevation of about -12 feet MSL, a five day setup period was put into effect. At this time in the grouting, a total of slightly over 21,000 CY of grout mix had been pumped, of which approximately 300 yards was OR-13, the unsanded mix, and the remaining quantity was OR-5, the sanded mix. When grouting operations were resumed, OR-5 was pumped into the B and C lines in gate bays 8, 9, and 10 until the grout level in all the holes in 8, 9, and 10, except 8 C', reached elevations varying between -6 and -9 feet MSL, or 1 to 4 feet below the base of the concrete slab. The elevation in 8 C' was -12 feet MSL, 7 feet below the base. Grouting operations were shifted and the D line behind gates 8, 9, and 10 was topped off into

the slab with OR-23. Grouting was once again resumed in the B and C line

of gates 8, 9, and 10, using the OR-23 topping out grout, and continued

until the level of the grout in all B and C borings in 8, 9, and 10 reached

an elevation of the top of the foundation slab.

After a considerable period of monitoring, numerous check borings were

made to depths corresponding to the depth of the underlying foundation

material. In each check boring, the grout material was continuous from the

base of the slab to the base of the original cavity, that is, the original

contact point between the base of the muck and the top of firm foundation

material. No muck was detected in any of these borings, indicating apparent

success in incorporating the muck into the original OR-13 mix. At this time,

the center of the drilling and grouting operations was shifted to gate bays

6 and 7 (Fig. 21), where initial borings had indicated a small cavity. Ad-

ditional borings made into gate bays 6 and 7 indicated a small cavity existed

along the B and C lines beneath the structure and behind the sheet pile along

the D line. (Figs. 13, 14, & 15). Since the size of the cavities in the B,

C, and D lines of gate bays 6 and 7 were shallow (a maximum of 7 feet in 7 C''),

it was determined to grout with OR-23 mix, and grouting was initiated in the

first part of February 1974. Initial grouting into 7 C and 7 C'' failed to

raise the level of grout in both areas and as a result, about 15 CF of foam

rubber material, followed by 31 CY of DOC mix, was pumped into 7 C. This

was immediately followed by OR-23 mix which began to rise until the grout

level in B, C, and D lines of gate bays 6 and 7 reached the top of the con-

crete slab. Check borings were made and indicated that grout material was intact from the base of the concrete slab to the base of the original cavity. With verification of adequate grout levels in gate bays 6, 7, 8, 9, and 10, the emergency grouting program at the Low Sill Structure was essentially completed. Additional check type borings are being drilled into gate bays 5 and 11, and, if necessary, grouting with OR-23 will be accomplished.

(Fig. 22) In conjunction with the grouting operations at the Low Sill Structure, the drainage system beneath the stilling basin and the forebay area was investigated. The investigation indicated the extreme downstream pipe No. 5 beneath the stilling basin to be intact and continuous. However, drainage pipe No. 3, located beneath the stilling basin and immediately downstream of the D line of borings, was found to be apparently ruptured and not continuous. An investigation of drainage pipes Nos. 1 and 2 in the forebay area indicated that both pipes were apparently ruptured and not continuous across the structure. Based on the results of these drainage pipe investigations, approximately 75 percent of the drainage system beneath the stilling basin is apparently intact, the exception being in the area immediately downstream of the sheet pile behind gate bays 8, 9, and 10.

(Fig. 23) In conclusion, it is considered that the successful emergency grouting operations at Low Sill has accomplished the following:

1. Defined the size of the original cavity and filled this cavity with suitable grout.

2. Given immediate lateral support to the steel batter and support piles.

3. Effectively sealed off the zone beneath the structure, particularly along the sheet pile between the structure and stilling basin so

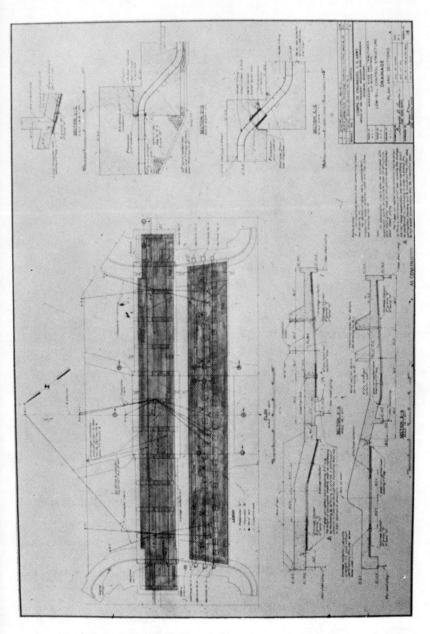

Figure 22

Figure 23

that no excessive uplift pressures are being measured anywhere.

4. Defined the potential efficiency of the drainage system in the
stilling basin.

ADDENDUM:

Since this paper was presented on 20 March 1974, three additional
borings have been completed in gate bay 5, and no cavity was found. In
addition, five borings have been drilled in gate bay 11, and a cavity
varying in size from 1.9 feet in 11 C'' to 34.5 in 11 B was found.
However, this cavity was not unexpected because of the early opening of
gate bay 11 on 6 December before the level of the grout could be brought
up into the base slab. This early opening of gate bay 11 was necessary
to prevent excessive heads from developing. The cavity was successfully
grouted with OR-23, and gate bay 11 was opened on 12 April 1974. One
boring has been drilled in gate bay 1 and no cavity has been found. It
is anticipated to continue drilling several holes in gate bays 2 and 3,
and, if necessary, to grout with OR-23. Continual surveillance will be
maintained to insure that the grout remains in place and that the stability of the Low Sill Structure is maintained. In addition, new piezometers will be installed in the stilling basin, and additional drainage
relief beneath the stilling basin will be provided for.

SPILLWAY FOUNDATION DAMAGE AND REPAIRS

BY: Andrew Eberhardt, Chief Structural

Engineer, Harza Engineering Company

Being a structural engineer I learned about foundations of
dams (whatever I know) by - exposure including the exposure to
geologists. Originally, I used to pity them. They would keep
poking around seemingly forever asking for additional core holes,
pits , adits, exploratory tunnels, etc. I would say: "Look, we
already drilled 3,000 feet. It is good rock." The answer would
be: "Well, there may be a horizontal, narrow clay seam down there
which we missed." You know, shadows of Malpasset Dam.*) Or the
answer would be: "I don't know what the company's policy will
be in this case, but I would not go along with what you recommend"
these sort of things. Very cautious, pessimistic, unwilling to
commit themselves people. However, after working on dams and particularly old dams for a number of years, I began to understand
geologists. Foundations can spring some surprises even after the
dam was built and put in operation. In my experience some soft
foundations performed better than some hard rock formations. Their
performance, was simply a function of the quality of the design.
Ultimately I concluded that the geologist is not so much concerned
about the foundation itself, but rather about what we, the engineers
will do to his foundation. This is what I am going to try to illustrate.

An example of good performance (considering the length of
service) is this small dam in Ann Arbor, Michigan (Fig. 1). Built
way back in 1914. Only 25 ft. high but of multiple barrel or arch
and buttress construction. An elegant design as the French say.
A delicate structure, 200 ft. long, without any expansion or contraction joints, resting on spread footings on river sands and gravels,
underlain by clay. After about 55 years of service, it did not display
any cracks or differential settlements. The original drawings showed
numerous piles under the footings. But each drawing also displayed
a brief note: "No piles were used." The stilling basin did not do .

---

*) It was thought at that time that such a seam lead to the known
   catastrophe.

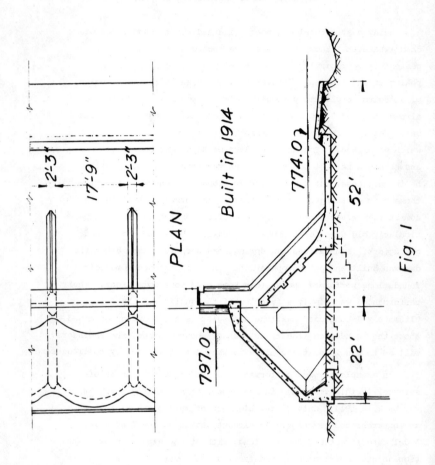

PLAN

Built in 1914

2'-3"

17'-9"

2'-3"

797.0

774.0

52'

22'

Fig. 1

as well. There was some undercutting at the downstream edge due
to the lack of rip rap protection. We also found extensive cavities
under the slab, but not over 3 inches deep. There was also a crack
in the stilling basin still. The crack was up to 2 inches open in
some places and up to 20 feet long.

The crack and the lack of rip rap and a filter must have lead
to the migration of the material, formation of cavities and scour.
How the crack occurred, we don't know. It could have been a poor
construction joint and defective concrete around it as it happens
often in old structures.

The repairs consisted of placing a concrete bulkhead, sealing
the crack, low pressure grouting, and placing rip rap and filter
material (Fig. 2). Considering the length of service and the causes
of the damage the soft foundation performed well.

Another dam stood on hard shale (Fig. 3). Hard shale, of course,
is a stronger foundation material than sands and gravels. A geologist
could even call it "competent" rock. (I am not sure whether an object
can be called "competent". One does not say "this bridge is competent."
This adjective is usually reserved for human beings. I suspect that
geologists identify themselves with rocks or consider them something
living. They discuss and argue their creation, changes, decay, etc.).
The foundation took some beating in the form of scour. This again was
not the fault of the shale. It was caused by lack of protection:
too short an apron without a sill or end deflector. In addition,
the operation of the gates throughout the years was highly improper.

The spillway had 10 tainter gates operated by a single carriage
hoist. The hoist had to be pushed from one bay to another with a
crow bar. The caretaker, an old man, was not willing to do it. So
he kept the hoist attached to one gate and would not move it to open
other gates until the first gate was fully open. As a result, every-
time there was a minor flood there, one or two gates would be fully
open at one end of the spillway. The tailrace was just one churning
powerful horizontal eddy. The result was nearly 12 ft. deep scour
which is quite a lot for a small dam.

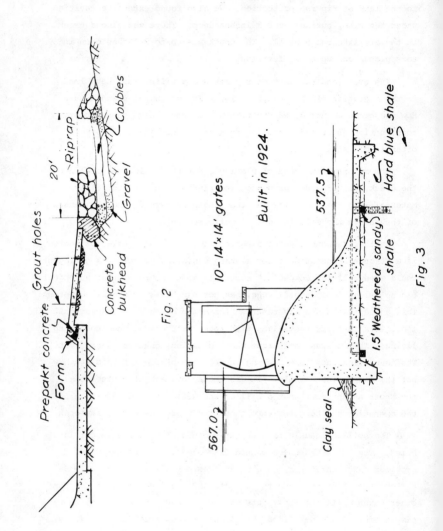

Fig. 2

Fig. 3

We unwatered the tailrace in 1951 to determine the exact extent of damage but found no undercutting of the concrete (Fig. 4). To save money, the repairs were limited to protecting and buttressing the toe of the apron with a concrete block anchored into rock (Fig. 5). We figured that if we can get the owners to operate the gates properly, there should be no additional scour. An electric motor drive was added to the hoist carriage for moving it from one gate to another. Shortly after these improvements were made, I stopped by to see how things were going. There was no change. A single gate at one end of the spillway was wide open and the tailrace was again one big and violent eddy. The caretaker still had to hook the chains to the hoist, then unhook and dog them. He thought this was still too much work. In a sense he was right. The current practice is to provide individual gate hoists.

But in 1972 or 20 years later a diver inspected the toe of the dam and the tailrace. He found everything in good order. There was no further scour and the new concrete was not damaged.

In my encounters with professional divers I learned to respect them. They all were intelligent, knowledgeable and alert people. Perhaps this is due to the natural process of elimination in this line of work. This cannot be said about some other professions.

Another experience was somewhat of a different nature. This was a low head powerhouse (35 ft. head). An old structure built around 1905 (Fig. 6). 15 years ago three of its units were converted into low level sluices. This is why the powerhouse qualifies for my talk. At that time hydraulic tests were made to determine the coefficient of discharge thru the sluice gates. But apparently nobody was concerned about the effect of this conversion on the powerhouse foundation. Two years ago, however, a diver sent down reported scour. This scared the owner who decided to unwater the draft tubes for inspection.

The unwatering was a costly operation because there were no bulk-heads or draft tube gates and no draft tube deck. We had to design a steel bulkhead 30' wide and about 25' high, built in three sections. A construction crane had to be brought in on a steel barge.

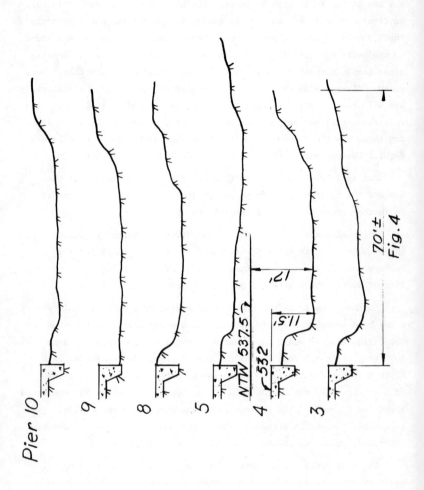

Pier 10   9   8   5   4   3

NTW 537.5

532

12'

11.5'

70' ±

Fig. 4

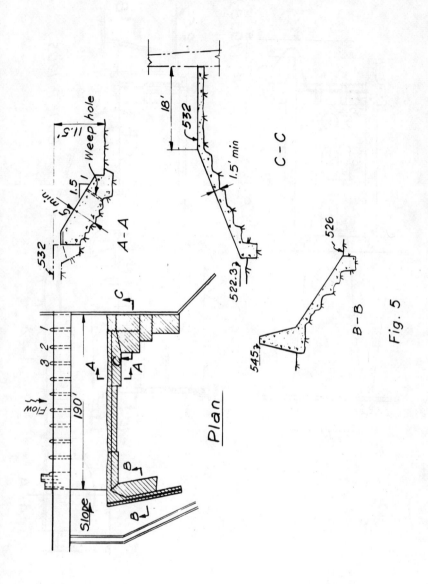

Fig. 5

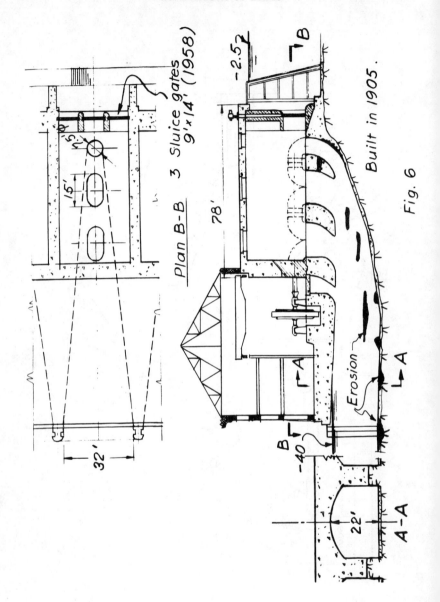

Plan B-B   3 Sluice gates
9'×14' (1958)

Built in 1905.

Fig. 6

Erosion

A-A

After unwatering we really did not find any significant damage
to the foundation. An interesting discovery was exposed rock in the
bottom portion of some of the piers. Up to 5 feet of the pier was
rock. Good sound dolomite. It was jointed but it suffered little
damage. Velocities in the draft tube were not high enough (10 fps.
average velocity).

Water pouring thru the intake floor openings would fall down
into the pool of water below. That is where most of the energy dis-
sipation took place. There were some occasional deep pockets of eroded
concrete. The damage, was very inconsistent from one draft tube to
another. Apparently it occurred wherever concrete was of poor quality
due to poor mixing and placement methods common nearly 70 years ago.

Actually, as it turned out, the critical moment was unwatering
itself. I was concerned about a blow out in the rock portion of
the piers under full head of T.W. in the adjoining draft tubes.
Fortunately, nothing happened.

The next example is a water supply dam founded on clay. The
dam was built about 1920. The spillway section about 500 ft. long
is supported on piles (Fig. 7). This practice has been largely
abandoned now. Some cavities were found under the apron in 1949.
They were filled with concrete and an end sill or lip was added to
reduce tendency for scour to undercut the apron.

In 1953 we drilled the apron and found extensive "roofing"
up to 3 feet deep.

It took 14,000 cu. ft. of grout to fill in the cavities. We
also added a steel sheet pile cut off at the apron end and rip rap
protection. The reservoir was also raised 5 feet by installing bascule
gates on top of the crest.

Cavities under a spillway built on piles can be caused by
several conditions: seepage under the dam, flow over the dam, eddies
caused by opening flood gates. Even a small separation of the foundation material from the underside of the structure opens a shortcut to
seepage. It is also possible that the original rip rap protection was
inadequate.

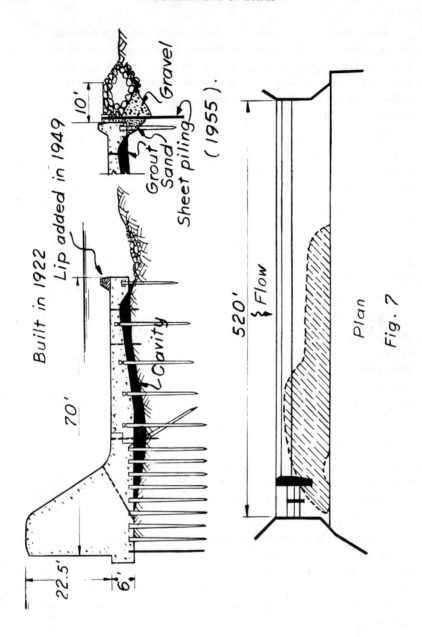

Built in 1922

Lip added in 1949

70'

22.5'

6'

Cavity

Gravel

Grout

Sand

Sheet piling

(1955).

520'

Flow

Plan

Fig. 7

So again: the foundation develops trouble because of the imperfection in the design of the structures.

A more spectacular foundation damage occurred in a dam in Virginia.

The dam itself is nothing very spectacular. It is a concrete gravity free overflow weir 70 ft. high and it is 520 ft. long (Fig. 8). In addition, there is an intake structure and a small powerhouse at the right bank. Then a short non overflow gravity section at the right end. The main purpose of the structure is to store water for some of Washington, D.C. suburbs.

Our report read: "the dam is underlain by granite gueiss of undetermined age (probable Paleozoic or Pre-cambrian)." I always realized that age is a very important and delicate matter when it comes to dealing with women, but engineering significance of the age of a rock formation always eluded me. I made an effort, however, to gain some insight into the thinking of the geologists and picked up a book on geology. It stated there that rocks of Paleozoic era are 250 million years old and those of Pre-cambrian age are 1,500 million years old. I do not believe we engineers are capable of understanding the meaning of these numbers.

There were, however, two significant aspects from an engineers view point: the rock was hard, but it was jointed.

The rock mass was thoroughly intersected by three principal sets of joints:

1. Approximately normal to the stream valley, dipping steeply upstream.

2. Approximately parallel to the stream valley dipping steeply riverward on the left bank.

3. Approximately normal to the valley dipping gently downstream.

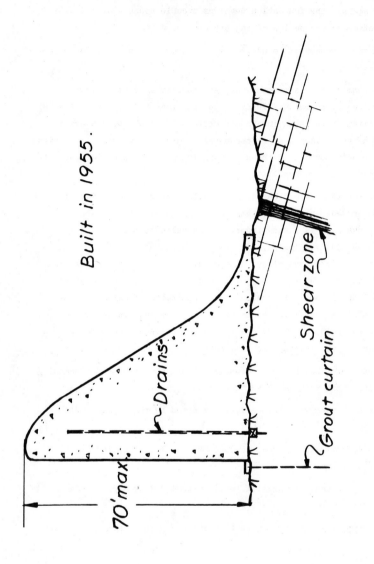

Built in 1955.

70' max.

Drains

Shear zone

Grout curtain

Fig. 8

The last set of joints represented danger of potential scour as correctly predicted by our geologist. There was also a shear zone a few feet wide which dipped steeply upstream and nearly paralleled the toe of the dam. The shear zone was tight and unweathered.

The dam has been in existence for about 20 years and a number of floods passed over it causing only a few feet of scour at the toe.

The 1972 flood (hurricane Agnes), however, more than doubled the maximum flow of 35 year record. The flood overtopped the intake structure, and the non-overflow section, flooded the powerhouse and ripped out the pipelines downstream.

The hard rock foundation proved very vulnerable. The water plucked out large blocks of rock like loose bricks. It was necessary to unwater the dam toe in order to assess the damage.

There were three main areas of scour as shown in Fig. 9: one 15 ft. deep at the right end of the spillway, another 20 ft. deep at the left abutment and the third up the abutment itself where a large volume of rock was removed by the flood. In the middle of the stream bed the scour was not significant except for the shear zone, indicating much tighter joints in the rock formation. The toe of the dam was slightly undermined in a few places where the shear zone approached or penetrated the dam foundation.

It became obvious that the main contributing factor to the damage (other than the joints in the rock) was poor hydraulic design: the lack of a concrete apron and sill to protect the rock and the lack of any provisions to lead away the water pouring over the left end of the dam.

The flood discharge of this portion of the dam hit the vertical cliffs of the abutment head on and was deflected laterally. As a result, there were flows from two directions: parallel and transverse to the river colliding over the left edge of the stream. This turbulence, lead to the deepest and the most extensive scour in this very area.

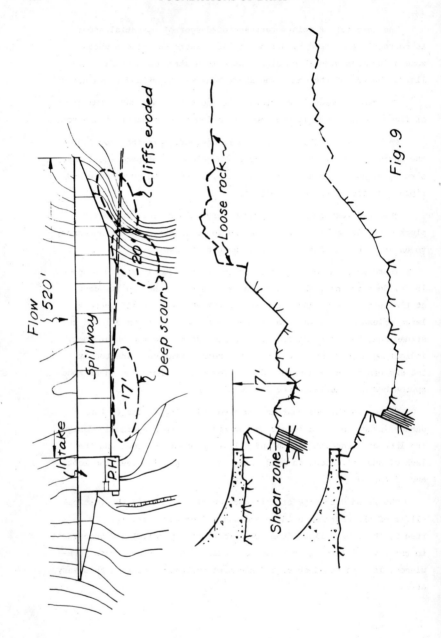

Fig. 9

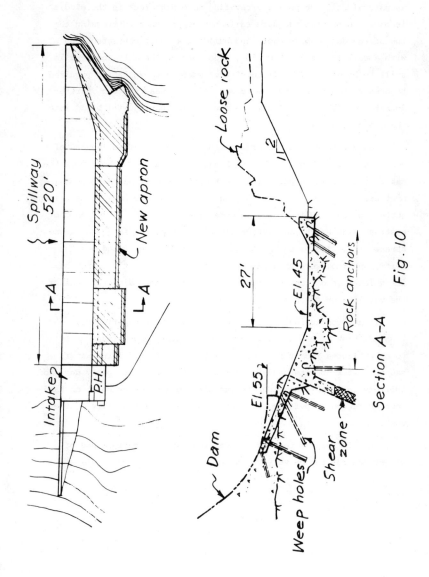

Spillway 520'

New apron

A

A

Intake

P.H.

Loose rock

27'

El. 45

Rock anchors

El. 55

Dam

Weep holes

Shear zone

Section A-A   Fig. 10

The remedial measures consisted of retaining the scoured areas as natural stilling pools, excavating some high rock in the middle in order to provide adequate tailwater depth, then buttressing the toe of the dam with concrete and building a short concrete apron with a deflector sill (Fig. 10). The work on the left abutment is still to be done. It will consist of some rock excavation required to open up the constructed waterway and to spread the flow farther downstream. The channel will be paved. The paving will blend into the toe of the weir.

One very clear lesson of the flood was great effectiveness of concrete paving in protecting rock against scour. The bottom of the narrow channel which parallels the toe of the weir on the left abutment was paved originally. The channel was not much more than a large gutter not over 10 feet wide in some places. It was formed on the upstream side by the dam (or weir) and by sheer cliffs on the downstream side. The paved bottom was untouched by the flood but the cliffs and rock were sheared off or plucked out exactly along the irregular edge of concrete paving which was placed directly against the vertical or steep cliffs.

If any conclusions could be drawn from these experiences, I would think of these: obviously any spillway foundation has to be protected from the forces of water: whether falling water, seeping water or water eddying horizontally. But a properly designed spillway on soft foundation may out perform a spillway built on hard rock where too much faith was placed in the invulnerability of the rock.

In this respect the lessons derived from the performance of old structures can be very valuable.

ENGINEERING GEOLOGIST'S RESPONSIBILITIES

IN DAM FOUNDATION STUDIES

by

Don U. Deere*

INTRODUCTION

The layout and design of nearly every element of an hydroelectric
project require close cooperation between the design engineer and the
engineering geologist. This is true for the cofferdams, diversion tunnel,
power tunnel, and surface or underground powerhouse, and even moreso for
the dam and its appurtenant structures -- the power intake and the spillway. Decisions must be made which will seriously affect not only the
time and cost of construction but the safety of the structures themselves.

At the present state of the art no one has to be convinced that the
details of the site geology must be known before reasonable layouts can
be made of the project elements, and before preliminary designs and cost
estimates can be developed in a truly realistic fashion. The factors of
(1) availability of natural construction materials, (2) the configuration
of the valley walls and channel, and (3) the foundation characteristics
clearly dictate the acceptability and relative suitability of dams of
different types and of different intake and spillway arrangements -- and
these factors are all geological.

The previous papers given at this conference clearly show that designers and contractors are able to cope with very major adverse geological

* President, Don U. Deere and Andrew H. Merritt, Inc., Consultants
  in Engineering Geology and Applied Rock Mechanics, Gainesville,
  Florida; and Visiting Professor, University of Florida.

conditions. However, the adverse conditions must be known in advance
so that the design, the specifications, and the construction planning
can take them into account sufficiently. If not, extensive delays, cost
over-runs, and controversy will certainly develop. If a latent adverse
geological feature remains undetected during both the design and construction phases, the potential of failure during operation in the following
years remains.

THE ROLE OF THE ENGINEERING GEOLOGIST

Feasibility Studies

The engineering geologist possibly may play his most important role
in the early days of the site selection and feasibility studies. It is
then that the general geology is being examined and major adverse features
such as regional faults, deep weathering, karstic limestone, old landslides, etc., are identified. The worse areas are eliminated, or at least
down-graded for the moment, and the better sites are studied in more detail.

Field reconnaisance with the aid of air photos is the major exploration tool. Test pits and trenches are helpful in examining the soil cover
and the weathered rock. Geophysical surveying is also used during this
phase in many cases -- usually seismic but occasionally geo-electrical.
Exploration galleries into the abutments and core borings may be done to
a very limited extent.

The end product of this phase of exploration is a preliminary geological
cross section of the valley showing the topography, soil cover, depth of
weathering, channel configuration and deposits, and rock structure. A geological map of the area is included with the geological cross section in an
Engineering Geology Feasibility Report for the site. The report includes
a description of the regional geologic setting, the site geology, foundation
conditions, and the availability of construction materials (clay, filter
and concrete sand, and rock for rockfill, concrete aggregate, and riprap.)
Foundation treatment (excavation depths and grouting) and suitability of
different dam types are discussed.

The various potential dam sites are compared and recommendations are
made for additional exploration for the one or two better looking sites.

Preliminary and Final Design Studies

The exploratory program will include many core borings and numerous exploration galleries as well as additional trenches and test pits. It is best to lay out the boring program not on a geometric basis but in a way so as to best furnish data both for unraveling the geologic framework at the site and for giving early information on the rock conditions important to the design engineer.

Some of the exploration may appear to be strictly for a geologic reasons; yet, the geological data obtained may be necessary for interpreting the site geology which, in turn, will often allow major benefits to be achieved in defining the best axis location, the required depth of excavation, the necessity of grouting, or perhaps the existance of a potential sliding problem.

The engineering geologist must be in constant communication with the design engineer throughout the exploration period. Potential problem areas are discussed and evaluated. Soil and rock mechanics tests may be planned; groundwater pumping tests, grouting tests, additional geophysical studies in the galleries, etc., must be considered and decided upon. All such testing must be carefully coordinated by the engineering geologist so that geologically representative areas are tested. Otherwise, interpretation of the data may be difficult and extrapolation of the data across the site may prove impossible.

Construction Phase

The presence of the engineering geologist during construction is paramount. The exposed rock must be evaluated, the design assumptions checked, and any newly discovered fault, spring, cave, deep weathering, etc., brought to the attention of the designers for any possible adverse implications. Grouting and the emplacement of drain holes can best be done when layed out in accordance with predominant joints or shear zones. The engineering geologist can be of help in laying out and monitoring such a program.

For all of the duties given above, it is assumed that the engineering geologist is familiar with general construction practices of excavation,

grouting, drilling drain holes, cofferdam construction, dewatering, fill
placement and rock treatment as well as with the geotechnical design
involving soil and rock mechanics. He may not be an expert in all these
fields, but he has a general understanding of them -- just as the design
engineer and geotechnical engineer have a general knowledge of engineering
geology. Communication is no longer a problem and mutual respect for
each other's fields is now the norm in dam studies.

SPECIAL ADVERSE GEOLOGIC FEATURES

It has rightly been said that the engineering geologist is not responsible for the presence of adverse geology at a site but that he is
responsible for finding out that such adverse features do occur. Thus,
the presence of buried peat, of an old landslide, of a buried channel,
of a solution cave, or of an important fault are all his responsibility
for searching out and identifying. And, these, he normally would discover
by his geologic mapping, air photo studies, borings, trenches, and galleries.

There are two features, however, that are not easy to detect and, yet,
are ubiquitous and treacherous. These are the foliation shear in metamorphic rocks and the shale mylonite seam in sedimentary rocks.

The Foliation Shear Zone

The foliation shear zone has been responsible for a multitude of design
and construction problems on hydroelectric developments constructed in metamorphic rocks.[1] The origin of the shearing is differential movements
between adjacent layers of metamorphic rock in geologic past caused by folding, or in some cases by stress relief associated with erosion and valley
cutting. The movements are concentrated in the weaker layers of the rock
mass, typically mica, chlorite, talc, or graphite schist in a sequence
of harder, massive rocks such as granite gneiss and quartzite. The movements are considered to be small, probably from a few inches to a few
feet (offsets of 5 ft. in pegmatite dikes were seen in one project in
Washington, D.C., and in another project in New York City).

----

[1] Deere, Don U. "The Foliation Shear Zone - An Adverse Engineering
Geologic Feature of Metamorphic Rocks," Jour. Boston Soc. of Civil
Engineers, Vol. 60, No. 4, October 1973, pp. 163-176.

The thickness of the shear zones is typically just a few inches counting the gouge and crushed rock. The adjacent rock, however, is heavily jointed and occasionally altered and slightly sheared for a few feet on each side. The shear zones are continuous over several hundred feet.

In the Washington, D. C., Metro Station (DuPont Circle) nine of these shear zones were encountered on a spacing of about 25 ft. In the very extensive water supply tunnels in New York City (City Tunnel No. 3 now under construction) the zones are only a few inches thick but they are sub-parallel to the tunnels and one zone may follow a tunnel for 500-700 ft. Such zones require steel sets for stability. The foliation shear zones are on spacings of 500-1000 ft. there.

At several underground powerhouses foliation shears have been met and conquered at a price (Churchill Falls, Oroville, Morrow Point)[2]. The Kariba arch dam in Rhodesia required extensive mining out and back-filling with concrete of a zone of crushed mica; a recent arch dam in Venezuela required extensive study, some pre-stressing of a foliation shear zone by a concrete slab and anchored tendons, and a minor redesign at the top of the arch.

The foliation shear zones creates a directional weakness in the rock in which the shearing resistance is entirely due to the residual friction angle of the crushed material, which in turn is a function of the grain size and mineralogy (values are in the general range of 15°- 25° but may be lower).

It behooves the engineering geologist to suspect that he will find shear zones parallel to the foliation, and to discuss with the designers the possible effects that these might have on sliding resistance, abut-ment stability, and foundation treatment. The orientation is critical, and this may be assumed to be parallel to the foliation. The actual location is also critical. Borings often do not recover the sheared, crushed rock and gouge. Exploration galleries or test shafts have proven to be the most definitive method of checking on the possible occurrence of these zones.

---

[2] Op. cit.

The writer has encountered critical foliation shears on more than
a dozen projects in the past few years. Remedial measures and redesign
have been necessary and total costs have been increased by many millions
of dollars. There is little doubt but that the foliation shear is truly
a "significant engineering geology feature."

The Shale Mylonite Seam

The shale mylonite seam is nothing more than a bedding plane shear
zone. Like the foliation shear, it is caused by differential movement
between adjacent beds (sedimentary rocks in this case) due to folding or
stress relief. The shearing is concentrated in the weak bed -- usually
shale but occasionally a thin bentonite or lignite bed -- which is bounded
by harder limestone, sandstone, or siltstone beds.

The sheared and crush shale gouge is often referred to in engineering
geology literature as shale mylonite (meaning ground-up shale). The zone
is usually only an inch or so thick but it may be continuous for several
hundred feet. The shearing resistance is close to the residual friction
value of 10°-20°, or lower depending upon the percent clay-size material
and the type of clay mineral.

Obviously, the sliding resistance of a concrete gravity dam could
be adversely affected by the occurrence of a near-horizontal mylonite
seam in the foundation work, as would an arch dam with a critically located
seam in the abutment. Cofferdams and lock walls founded above such seams
have also failed in a couple of instances. There is no doubt but that the
shale mylonite seam is also "a significant engineering geology feature".

STABILITY OF CONCRETE GRAVITY INTAKE STRUCTURES

The sliding stability of a concrete gravity dam is always checked
in design and is usually found to be sufficient -- often because of its
low position in the valley cross section. There is another class of structures, however, that is more common than normal gravity dams and often
more difficult to establish stability. These are the intake structures.

Concrete gravity intake structures are used with all types of dams
including fill dams. The structures may be power intakes or spillway

headworks structures. They are commonly located at one end of the dam, often rather precariously perched on a cut surface in the valley wall.

Stability of the rock slope above the spillway may be a problem, but of greatest concern is the sliding resistance of the gravity structure under the hydrostatic thrust of the reservoir water. This is particularly critical when the topography immediately downstream falls off rapidly to the valley bottom below, and when the intake structures are moved as far as possible downstream so as to reduce the length of the spillway chute or penstocks.

Even in normal gravity dams or low fill dams with the power intake located in the bottom of the valley, there is the tendency to found the concrete gravity intake at a higher elevation than the powerhouse immediately downstream with a prominent rock slope between them. Occasionally, the intake and powerhouse will be designed as a structured unit; often, they are not and the stability of the intake structure perched on a rock ledge becomes critical.

For the above cases, any adversely oriented geologic feature (near-horizontal) that would be located beneath the foundation would be critical--particularly, if the feature is weak and if it is continuous so as to day-light on the downstream natural slope or cut slope. Obviously, the previously described foliation shears in metamorphic rocks and shale mylonites in sedimentary rocks could form such weak and continuous feature.

CONCLUSIONS

Careful exploratory work must be carried out in all phases of a dam project. Coordination with the dam designers is imperative throughout. The engineering geologist is responsible for determining the geologic framework at the site and in detecting the adverse features and discussing them with the designers.

Not only must the normally well understood geological features such as faults, deep weathering, solution channels, joints, etc., be studied and depicted, but the less well understood "foliation shear" and "shale mylonite" must be checked for. These features are common and "significant" because of their considerable continuity and weak strength. If critically oriented and located beneath concrete structures, they could bring about sliding failures.

The engineering geology of dam foundations must certainly include an assessment of these shear zone features. Because of their small thickness, very careful mapping, special drilling and sampling, and even exploration galleries must be used in order to detect these zones. It is a major part of the engineering geologists' responsibilities to see that this assessment is vigorously but meticulously done. Lack of time and lack of funds can not be an excuse; the responsibility can not be shed.

GEOTECHNICAL STUDIES FOR LA ANGOSTURA PROJECT, MEXICO

by Raúl J. Marsal*

INTRODUCTION

The design, construction and even the operation of a dam, require the combined efforts of the geologist and the engineer in order to establish a thorough understanding of the geotechnical problems involved.

In the past, geologists commonly worked in isolation and produced thick reports containing general descriptions of the regional topography and geology, some information on the reservoir area and the account of explorations undertaken at the damsite. There was little interaction, if any, between geologist and engineer and the result usually was that the engineer failed to take into account important facts revealed by the geological survey. On the other hand, the geologist knew little of the bearing of these features on the project as a whole or on any particular structure. Therefore, no special studies were performed in time to avert a condition unfavourable either from the standpoint of economics or safety. Fortunately this state of affairs has improved as we travelled along a lengthy and sometimes arduous path, but much remains to be done to coordinate the joint interests of engineers and geologists.

To illustrate several aspects of the type of cooperative effort required, the geotechnical studies performed by the Comision Federal de Electricidad (CFE) and the Instituto de Ingeniería, UNAM, at the design stage and during the construction of La Angostura Project will be examined.

---

*Research Professor, Instituto de Ingeniería, UNAM

La Angostura Project is a hydroelectric power plant located in the upper Grijalva River, State of Chiapas (Fig 1). The dam is of rockfill type, 145 m high; it has a central, inclined clay core and pervious shoulders made of rockfill, sand and gravel. The total storage capacity of the reservoir is 18 x 10^9 m^3, of which 7 x 10^9 are needed for flood control. The spillways, two gated open channels with ski-jumps located in the left bank (Fig 2), are designed for a maximum discharge of 6,000 m^3/sec. The powerhouse was built in a cavern (right bank, Fig 2), approximately 22 m wide, 40 m high and 100 m long; it contains 3 x 150 MW units and four power transformers. The turbines are fed by lined tunnels 6.5 m in diameter, connected to the intake works through shafts where the emergency and service gates are installed. The river was diverted by means of two cofferdams, the upstream one being 60 m and the downstream 30 m high (Fig 3), and two concrete lined tunnels, 13 m in diameter.

GEOLOGY

The upper Grijalva River flows along the Chiapas Plateau through La Angostura Canyon, located 700 km South-East of Mexico City. For a schematic presentation of the structural geology and tectonics of this region see the N-S section of Fig 4. The plateau is divided in several blocks due to thrust faulting and the crystaline basement is underlain to the South by the Pacific basaltic crust.

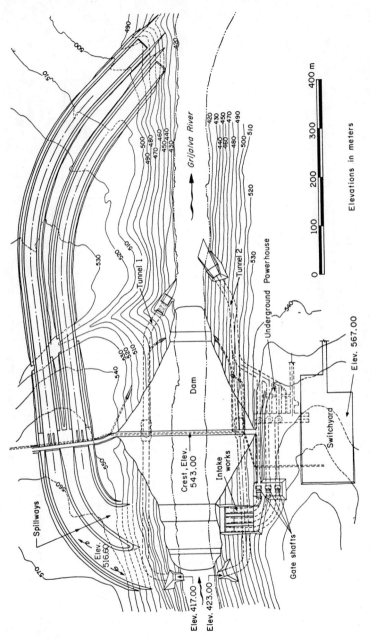

Fig 1. Plan view of La Angostura Project

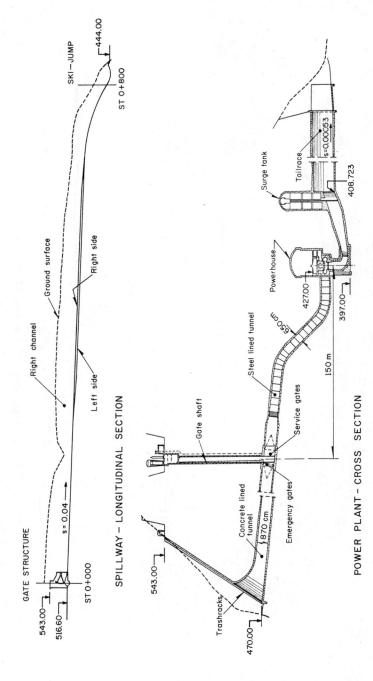

GATE STRUCTURE

543.00

516.60

ST 0+000

s = 0.04

Right channel

Ground surface

Right side

Left side

Right side

SKI–JUMP

444.00

ST 0+800

SPILLWAY – LONGITUDINAL SECTION

543.00

Trashracks

470.00

Concrete lined tunnel

870 cm

Emergency gates

Gate shaft

Service gates

Steel lined tunnel

650 cm

Surge tank

Tailrace

s=0.00053

408.723

Powerhouse

427.00

397.00

150 m

POWER PLANT – CROSS SECTION

Fig 2. Spillway and Power plant

① River bed materials
② Dumped rock
③ Dumped clayey soil
④ Compacted clayey soil
⑤ Compacted sand and gravel
⑥ Rockfill

Total volume : Upstream cofferdam 490,300 m³
Downstream cofferdam 101,000 m³

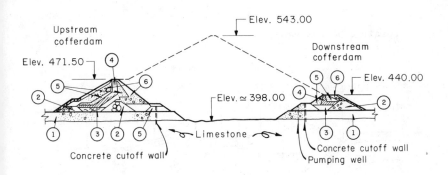

Fig 3 . Cofferdams for diversion of the river (Ramírez de Arellano and Moreno , 1971)

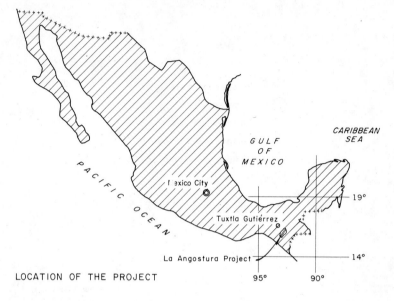

LOCATION OF THE PROJECT

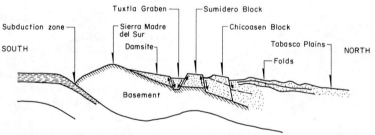

N-S SECTION OF CHIAPAS (F. MOOSER)

Fig 4 . Location of La Angostura Project and structural
            geology of Chiapas

In the reservoir area, the river cuts through
stratified sedimentary formations of the Jurassic and Cretaceous ages composed of limestones and shales. These strata
dip toward the North-East with an average inclination of
8 degrees.

Fig 5 presents the geology of the region bounded by the Grijalva and Santo Domingo rivers, both running in
an almost parallel northerly direction. The map indicates the
folding and the approximate boundaries of the different geological units that. were identified in field inspections and by
photogeological interpretation. The sequence of these units
in a N-S section is attached to the above figure. The highly
karstic stratum designated UD, underlies others of pure to
clayey limestone; at La Angostura, the latter are covered by
reefy formations and shales.

The structural geology and tectonics of this
area is presented in Fig 6. The set of fractures was further
traced and explored in the vicinity of the damsite, with the
purpose of evaluating seepage losses through the abutments
(see Fig 10).

Seismicity. Studies based on the location of epicenters and
intensities of earthquakes in the upper Grijalva basin, indicate that this region is very active from the standpoint of
seismicity. Since 1900 to date, 678 earthquakes of a magnitude exceeding 5 (Richter Scale) had the epicenters located in

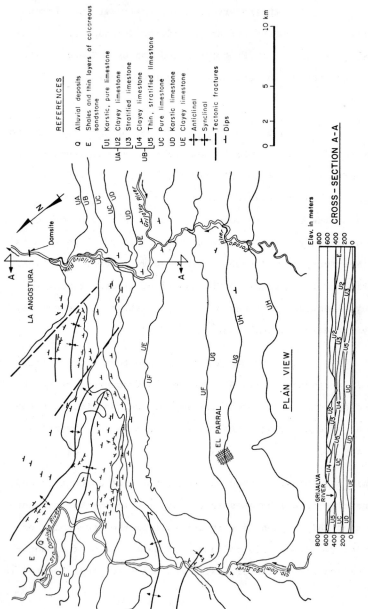

REFERENCES

Q   Alluvial deposits

E   Shales and thin layers of calcareous sandstone

UA   U1   Karstic, pure limestone
     U2   Clayey limestone
     U3   Stratified limestone
UB   U4   Clayey limestone
     U5   Thin, stratified limestone

UC   Pure limestone
UD   Karstic limestone
UE   Clayey limestone

✛   Anticlinal
✛   Synclinal
- - -   Tectonic fractures
—   Dips

0   2   5   10 km

PLAN VIEW

CROSS - SECTION A - A

Fig 5. Geology of the region bounded by the Grijalva and Santo Domingo rivers

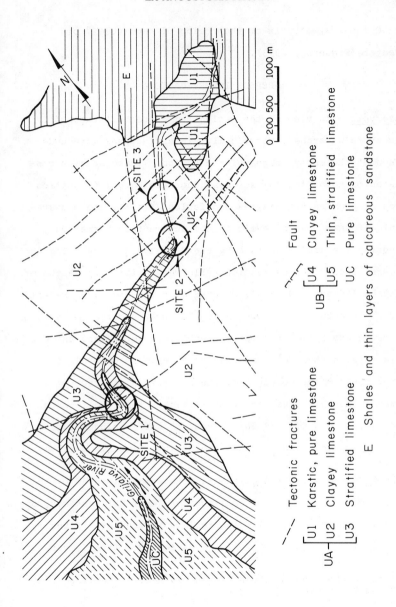

Fig 6. Geology and tectonics of La Angostura Canyon

Tectonic fractures

Fault

UA ⎧ U1 Karstic, pure limestone
    ⎨ U2 Clayey limestone
    ⎩ U3 Stratified limestone

UB ⎧ U4 Clayey limestone
    ⎩ U5 Thin, stratified limestone

UC Pure limestone

E Shales and thin layers of calcareous sandstone

0 200 500   1000 m

the Chiapas area*, six of them being of magnitude 7 or
greater (Figueroa, 1973).

SEEPAGE INVESTIGATIONS

Reservoir. In view of the geological characteristics of La
Angostura watershed, permeability investigations were undertaken at the early stages of the design. These studies consisted of topographical surveys, photogeological identifica-
tion of prominent fractures and faults (Fig 6), and measurements of water table levels in the sedimentary formations,
covering the area comprised by the Grijalva and Santo Domingo
rivers. About 50 observation wells were drilled to depths of
between 100 and 200 m, distributed over two areas, El Parral
and La Angostura, as shown in Fig 7. Over a period of several
years (since 1967), the water levels were determined and correlated with those of the Grijalva River. In general, the
water table rises gently from the river to a distance of 2 km
inwards, at which the elevation is higher that the intented
pool level (N.W.S., 539.50). Thus, it was concluded that a
seepage connection with the Santo Domingo River is unlikely.

Damsite. The selection of the damsite was influenced by the
presence of a highly karstic limestone (unit UD) that outcrops
in the reservoir (Fig 5). In preliminary studies, Sites I,
II and III shown in Fig 6 were investigated. Site I, very at-

*Region bounded by latitudes 14 and 19 degrees and longitudes
 90 and 95 degrees (see Fig 4).

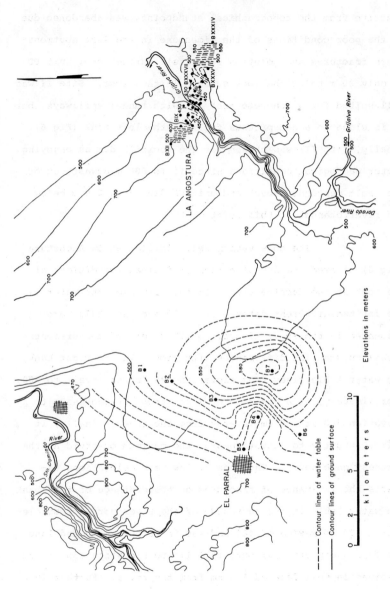

Fig 7. Water table observations in the Parral and La Angostura areas

----- Contour lines of water table

——— Contour lines of ground surface

Elevations in meters

kilometers

0   2   5   10

tractive from the topographical standpoint, was abandoned due
to the poor conditions of the limestone in the left abutment
(open fractures and solution conduits); furthermore, unit UD
is only 20 m below the rock surface in the river. Site II was
well suited for a concrete arch dam with tunnel spillways, but
it is close to a conspicuous fault in the left bank (Fig 6).
Finally, Site III was chosen for the rockfill dam as enjoying
better geological conditions since it is 600 m downstream of
the fault mentioned about and unit UD dips down 200 m below
the river channel at this point.

The observation wells drilled at La Angostura
(Fig 8) allowed the investigation of drainage conditions of
the rock at the damsite area. Contour lines of the water table for two different dates (Oct. 1970 and Jan. 1972) are
presented in the above figure. Neither sets of measurements
reveal substantial differences with time. In the right bank,
the water table rises from elevation 420 (river level) to 500,
approximately, at a distance of 1 km; then, it decreases to
elevation 460 at 1.5 km inwards from the river; finally, it
climbs abruptly to elevation 595, the maximum observed at the
farthest gage point. The zone of low water elevations reflects the influence of fractures on the drainage of this bank.
The water table in the left side of the river varies from elevation 420 to maxima of 485 and 450 in October 1970 and January 1972, respectively; but, it falls to elevation 435 in the
observation well located 1.5 km from the river. Further in-

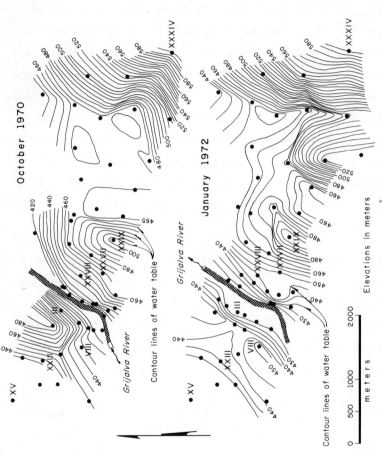

Fig 8. Water table observations at La Angostura

side this bank, the water table rises again and attains elevations above the pool level (N.W.S., 539.50).

Water Absorption Tests (WA). Tests of this type were performed in most of the borings drilled at the damsite either for
coring or piezometric observation. WA-values varied within a
range depicted by the examples given in Fig 9. Water absorptions of 10 to 20 Lugeons were observed at some test sections,
suggesting the presence of fractures or solution conduits in
the limestone. However, a tendency of WA to diminish with
depth was noted; with a few exceptions, WA was less than
2 Lugeons below river level. All the borings tested at this
stage of the geological exploration were vertical.

        The apparent inconsistency between water table
measurements and the results of the water absorption tests
commented above, promoted an additional study of the drainage
conditions of the rock by means of inclined borings drilled in
the vicinity of prominent fractures. Each of the selected
fractures was intersected by three borings at different depths
and water tested. The distribution and values of water absorption did not reveal much difference with those registered
in vertical holes. It was confirmed that one has to expect
greater permeability above the water table than in the deeper
strata of both banks.

        Based upon the above information, it was decid-
ed to 1) confine the grout treatment of the foundation and a-

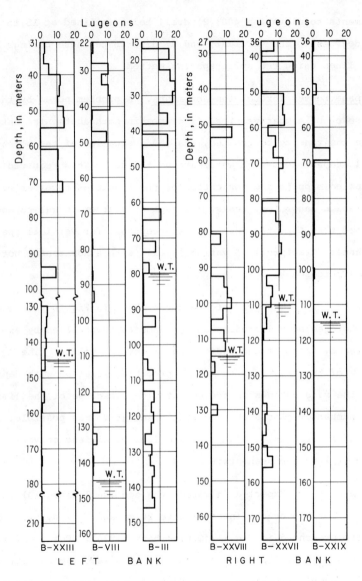

Fig 9. Water absorption tests at the damsite

butments to elevation 400; 2) drill holes inclined at 15 to
20 degrees to the vertical, and 3) distribute these borings in
two lines using the split-spacing method.

Seepage Losses Through the Abutments. A problem posed by the
presence of almost vertical fractures in both abutments, outside the zone to be grouted, was the evaluation of the potential seepage losses of water around the dam. This required a
detailed mapping and inspection of the fractures together with
hydraulic laboratory tests. Through the latter it was established that Dupuit's formula was acceptable for vertical open
channels (fractures) of variable width with a horizontal bottom (Cruickshank, 1970).

The grid of fractures shown in Fig 10 was based
on photogeological studies at the damsite, controlled by exploratory trenches and addits dug at several points. The
width of the fractures and their probable variation with depth
was the main unknown in the computations, because of the limited number of places where direct observation was possible.
Therefore, ranges of width ($\bar{w}$) and its coefficient of variation (V_w) were chosen arbitrarily.

Expected values of the seepage losses E(Q) in
m^3/sec for $\bar{w}$ over the interval 1 to 20 mm and V_w = 0, 0.5 and
3, are presented in Fig 11 (Cruickshank, 1970). For $\bar{w}$ = 5 mm
and 0 < V_w < 3, E(Q) varies from 1 to 10 m^3/sec, approximately. Economic studies of the effect of water losses in the en-

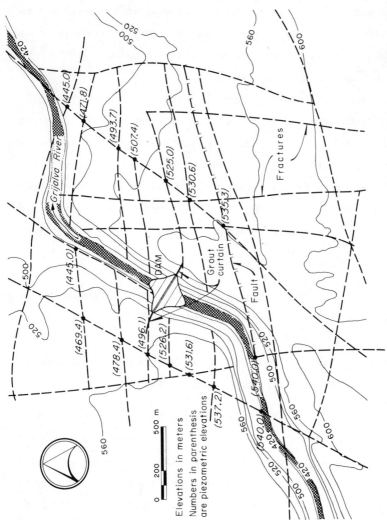

Fig.10. Grid of fractures and some of the computed piezometric elevations at La Angostura damsite

Elevations in meters
Numbers in parenthesis
are piezometric elevations

ergy production of the power plant, showed that a maximum seepage value of 20 m³/sec, about 7 per cent of the average rate
of flow of the river, was admisible. Hence, the decision was
taken 1) to limit the abutment treatment to about 100 m inside
both banks in order to take care of tectonic fractures as well
as those parallel to the canyon developed by stress relief,
and 2) to find a feasible and economic way to seal, close to
the slopes, the most prominent fractures connected to the reservoir. In this respect, a new series of exploratory trenches
and galleries in the vicinity of several fractures was excavated upstream of the damsite. Only the fault located on the
right bank (Fig 10) was grout treated, as described below.

ROCK MECHANICS STUDIES

For design purposes of the underground powerhouse and the stability investigation of spillway cuts, in
situ and laboratory determinations as well as field measurements during the excavation of these structures were undertak-
en by CFE and analized by J. Alberro (1973).

Results of laboratory tests performed with
specimens sampled in one of the three borings drilled around
the powerhouse, are presented in Table 1. Standard deviations
of the porosity, unconfined compressive and tensile strengths
and moduli of deformation, show a rather high scattering in
these characteristics for geological formations U2 and U3
(Fig 5), in which the cavern was to be excavated.

TABLE 1. TEST RESULTS FOR SAMPLES OF BORING CM−1 AT THE POWERHOUSE AREA

Test parameter	Formation	Number of tests	$\bar{x}$	σ
Porosity (n), %	U2	44	14.1	9.1
	U3	38	22.8	5.0
Unconfined compressive strength (σ_c), kg/cm²	U2	55	404	96
	U3	51	202	79
Tensile strength (σ_t), kg/cm² (Brazilian test)	U2	23	32.4	46
	U3	20	22.9	11
Modulus of deformation (D), 10³ kg/cm²	U2	27	166.3	59.3
	U3	29	126.5	46.7

$\bar{x}$=average value σ=standard deviation

TABLE 2. IN SITU MODULI OF DEFORMATION, GALLERIES 2 AND 3, LIMESTONE FORMATION U2 (Alberro, 1973)

Test site	Type of test	Direction	Number of tests	Mean value of D, in kg/cm²
Gallery 2 (in a zone of sound rock)	Rigid plate ϕ = 28 cm	//	4	130 380
		$\perp$	6	43 740
	Flexible plate ϕ = 1 m	//	4	126 830
		$\perp$	2	55 980
	Goodman jack	//	6	57 500
		$\perp$	2	49 000
	Micro-seismic	//	1	190 000
		$\perp$	1	150 000
Gallery 3 (close to a zone affected by vertical fractures)	Rigid plate ϕ = 28 cm	//	3	17 520
		$\perp$	4	40 760
	Flexible plate ϕ = 1 m	//	1	54 100
		$\perp$	2	44 840
	Microseismic	//	1	170 000

D = modulus of deformation

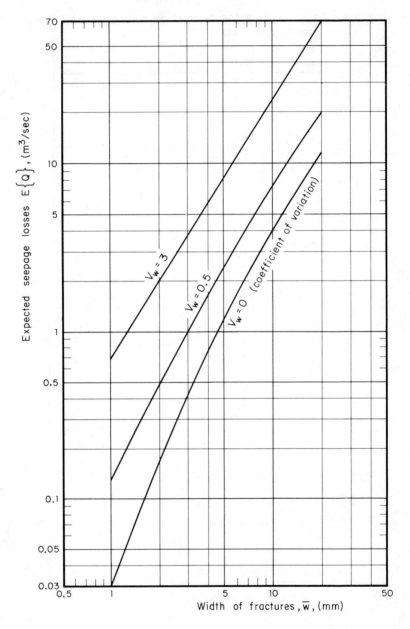

Fig 11. Seepage losses vs width of fractures (Cruickshank, 1970)

Table 2 presents average values of the static
and dynamic moduli of deformation determined in two galleries
dug in the right bank, 12 m about the vault and normal to the
longitudinal axis of the cavern. These tests were made with
rigid and flexible plates, the Goodman jack and the microseis-
mic method, perpendicular and parallel to the stratification.
Note the differences between the values of the static modulus
parallel to the stratification for Galleries 2 and 3, which
may be explained by the presence of fractures in the vicinity
of the latter test gallery. Also, the Goodman jack values are
low; this is attributed to fissuring around the test hole
caused by horizontal tectonic stresses. Based on 1) the dynamic results; 2) the frequency of transversal waves, and 3) a
known empirical correlation, the computed static moduli are
50,000 and 90,000 kg/cm^2 for the normal and parallel directions, respectively. These values agree with those obtained by
means of static load tests.

Stress relaxation tests were carried out in the
powerhouse area (Gallery 2), to investigate the magnitude and
direction of tectonic stresses prior to excavation. The results of these measurements, combined with the computation of
stresses due to the overburden, disclosed a horizontal tectonic stress parallel to the river of 80 kg/cm^2.

An example of the effect of blasting on the
slopes of the spillway channels is illustrated by Fig 12. The

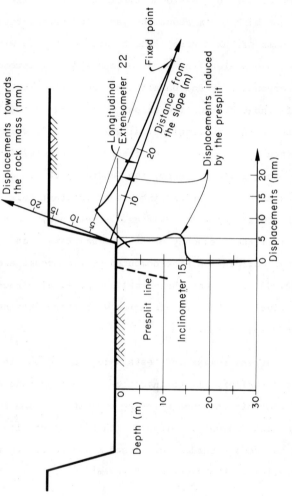

Fig 12. Displacements induced by presplit. Right spillway
channel, St. 0+050 (Alberro, 1973)

displacements induced by the explosives (pre-split) during the
excavation were measured with longitudinal extensometers and
inclinometers. The horizontal deformations detected by Inclinometer 15 (Fig 12) was of 5 mm, about constant for a depth of
15 m, and nil for the underlying strata. On the direction of
Extensometer 22, the displacements toward the hill increased
from zero to 6 mm in the first five meters from the slope, and
thereon decreased to zero over a length of 30 m.

Fig 13 shows measurements undertaken in Diversion Tunnel 1 during its excavation (performed in two stages)
and afterwards. Displacements toward the tunnel in Points 1
and 3 registered with longitudinal extensometers were 8 and
5 mm, respectively, at the end of the first stage of excavation; they increased linearly with time and the effect of the
second stage is not appreciable. Furthermore, the total vertical displacement of Point 3 was 13 mm over five months of
observation, which means a deferred deformation 1.4 times
greater than the initial deformation (5 mm).

Based on the mechanical characteristics of the
rock and the evaluation of tectonic stresses, Alberro (1973)
computed the deformations of the mass in the vicinity of the
cavern for the powerhouse using the finite-element method.
Fig 14 presents a comparison of computed and measured horizontal displacements at the end of excavation; the ratio between
them is about two. As commented by Alberro, to account for

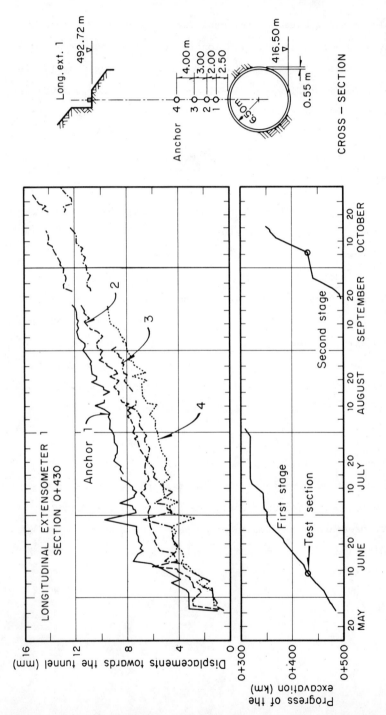

Fig 13. Vertical displacements in the rock upon excavation of Diversion Tunnel 1(Alberro,1973)

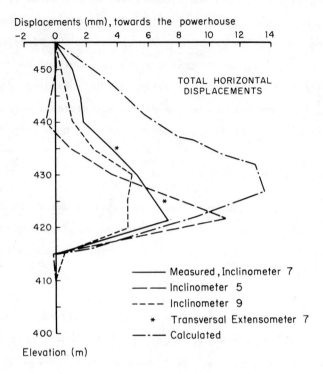

Fig 14. Total horizontal displacements at the
end of the excavation, measured and
calculated (linear elastic behavior)
(Alberro, 1973)

this difference one has to bear in mind 1) the effect of the
blasting on the deformation of the mass, as disclosed by observations in the spillway (Fig 12); 2) the deferred deformations of the rock revealed by measurements in the Diversion
Tunnel 1 (Fig 13), and 3) that the static moduli of deformation used in the analysis was determined in a loading process,
whereas the stresses around the cavern are decreasing upon
excavation. Had the dynamic moduli (Table 2) been used, the
correlation between measured and computed deformations would
have been almost perfect.

　　　　The above geotechnical studies helped in the
planning of the excavation and reinforcement of the cavern,
and also in the design of the gate structures of the spillway.

DAM CONSTRUCTION

Design. Feasibility studies which included several alternatives for the dam, powerhouse and spillway, showed that the
most convenient scheme was that composed of a rockfill dam,
two gated channels for the spillway and an underground powerhouse in the right bank (Figs 1 and 2).

　　　　The initially designed cross-section of the dam
is shown in Fig 15. A rather thin core of compacted soil in
the central portion supported by masses of rockfill, transitions zones and filters, are the main features of this section; the upstream outer slope is 2H:1V and the downstream,

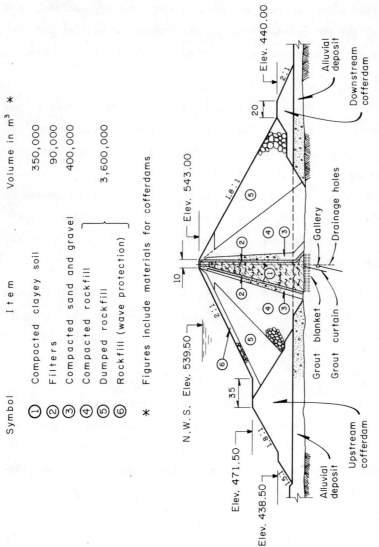

Symbol	Item	Volume in m³ *
①	Compacted clayey soil	350,000
②	Filters	90,000
③	Compacted sand and gravel	400,000
④	Compacted rockfill	
⑤	Dumped rockfill	3,600,000
⑥	Rockfill (wave protection)	

* Figures include materials for cofferdams

Fig 15. Maximum cross-section of the dam as originally designed

1.8H:1V. The cofferdams are incorporated in the dam and removal of the alluvial deposits in the river channel was pre-
scribed along the core and part of the pervious sections. A-
bout half the volume of the rockfill zones as well as the filters and transitions were to be compacted in layers by means
of vibratory rollers.

The use of about 75 per cent of the rock to be
excavated from the spillway area, was the economic concept of
this design. The abovementioned percentage was based on the
examination of cores obtained from borings drilled along the
original location of the spillway and the experience gained in
previous cuts along access roads. For reasons explained later, the estimation of useful rockfill was grossly in error and
caused significant changes in the cross-section of the dam.

Foundation Treatment. As indicated above (Fig 15), the design
called for the removal of the alluvial deposit prior to the
placement of the core and part of the pervious zones. Exploratory borings showed that this deposit was essentially composed of sands and gravels, but there were doubts about the
distribution of these soils and the possible existence of some
layers or lenses of silty materials. Excavations in the river-bed (Fig 16) disclosed a mixture of sand and gravel inter-
spersed with lenses of medium to fine sands; no silty or clayey soils were found.

Fig 17 shows the character of the rock upon

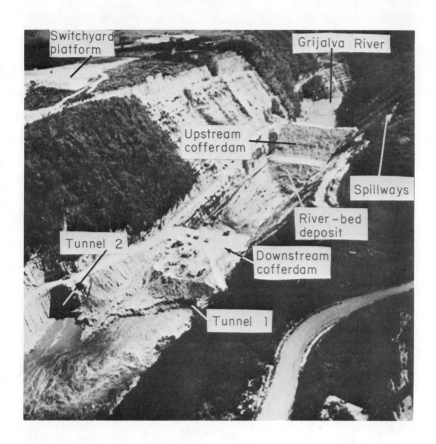

Fig 16. Excavation of river-bed deposit

Fig 17. Trimming and cleaning of the
rock foundation

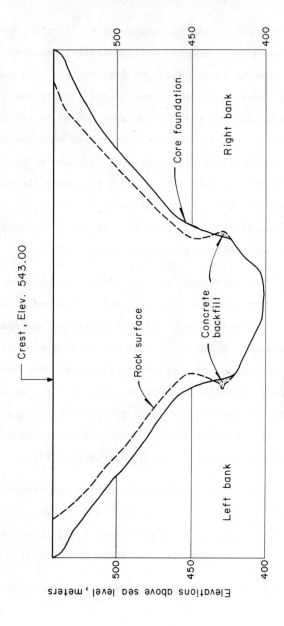

Fig 18 . Cross - section of La Angostura Canyon and profile of core foundation

cleaning and trimming of the exposed surface. At the bottom, the rock was sound and presented small potholes, smoothed fractures and some karstic zones. On both sides of the canyon, the stratification of the limestone was clearly revealed by layers of hard clay or marl, which were eroded by the river to a depth varying from 10 to 50 cm. Weathered rock was removed and the important defects corrected by means of dental work. It is worthwhile noting that, upon finishing the excavation between cofferdams, the total seepage through the abutments and the cutoff walls was less than 20 l/sec for water heads of 40 m (Ramírez de Arellano and Moreno, 1972).

The transversal profile of the canyon shown in Fig 18 reveals pronounced overhangs on both sides and from elevation 520 to the crest (elev. 543) the limestone was fractured and weathered. These conditions of the abutments required 1) the excavation of a trench to place the core over sound rock, and 2) trimming and concrete fillings to improve the shape of the foundation surface. The profile of this surface along the core was recommended upon studies of tension zones by the finite-element method (Covarrubias, 1970).

The entire core-rock foundation was treated with water-cement grouts injected through 2 in. borings 8m deep, under a pressure of 1 kg/cm^2. Grout takes of this blanket in different zones of the foundation are shown in Table 3.

Both the river channel and the abutments were

TABLE 3. DATA ON THE GROUT BLANKET

Section	Length of drilling (m)	Cement consumption (ton)	Grout takes (m³/m)			
			Stage I	II	III	IV
River channel	4 900	191	0.09	0.07	0.18	–
Right bank	5 690	611	0.32	0.21	0.28	0.24
Left bank	5 250	545	0.30	0.31	0.26	0.08

TABLE 4. DATA ON THE GROUT CURTAIN

Section	Length of drilling (m)	Cement consumption (ton)	Grout takes (m³/m)			
			Line A	B	C	D
River channel	2 030	64	0.14	0.05	–	–
Right abutment	2 260	190	0.26	0.13	0.24	0.34
Left abutment	1 970	102	0.16	0.14	0.23	0.05
Gallery No 1	6 910	74	0.06	0.03	0.02	–
Gallery No 1 bis	1 590	23	0.04	0.06	0.09	–
Gallery No 2	6 520	166	0.10	0.06	0.04	0.12
Gallery No 3	5 680	957	0.60	0.37	0.31	0.23
Connection No 3	2 800	732	1.72	0.61	0.17	0.10
Gallery No 4	7 880	701	0.31	0.20	0.11	0.18
Gallery No 5	9 300	205	0.10	0.06	0.06	0.04
Gallery No 6	1 900	2	0.02	0.04	–	–

grout treated down to elevations 360 and 400, respectively.
This work was done from galleries excavated at three elevations (Fig 19), one of them below the rock surface at the river channel (elev. 369). Mixtures with water-cement ratios of
1:1 to 10:1 by volume, were injected in 1.5 in. inclined holes
drilled in at least two lines following the split-spacing
method. A summary of this treatment is presented in Table 4.
The total length of drilling was 48,840 m and the corresponding cement consumption of 3,216 ton; the average grout takes
ranged from 0.03 to 1.73 m^3/m. From the abovementioned galleries and upon finishing the grouting operations, a series of
holes 3 in. in diameter at 10 m centers, were drilled to provide drainage.

After studies on seepage losses through fractures and subsequent exploration of the most important ones
connected to the reservoir, it was decided to treat the fault
located in the right bank 600 m upstream of the dam (Fig 10).
The other fractures explored pinched a few meters inside the
hill or had impervious fillings. The treatment of the abovementioned fault was performed from three galleries, using
grout mixtures injected through a series of overlapping borings that extended from pool level to a few meters below the
river deposit.

The effectiveness of the whole treatment described above is not known since the first filling of the reservoir will start on May 1974.

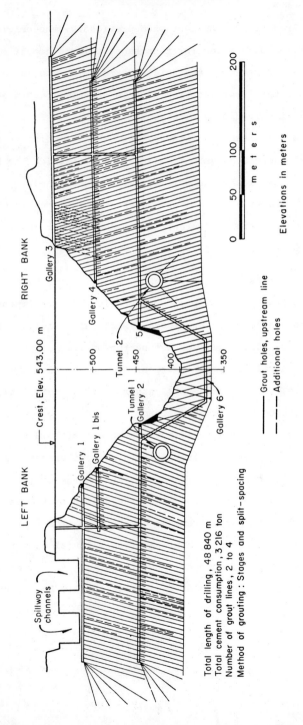

Total length of drilling, 48 840 m
Total cement consumption, 3 216 ton
Number of grout lines, 2 to 4
Method of grouting : Stages and split-spacing

——— Grout holes, upstream line
– – – Additional holes

Fig 19. Grout curtain

Rockfill Production. As mentioned previously, the original
design of the dam called for the use of a large amount of the
rock to be excavated at the spillway area. It should be mentioned that the first location of this structure was near the
canyon and involved open cuts up to 100 m high. Since this
was deemed undesirable in view of structural conditions of the
rock mass, the discharge channels were moved inside the hill
as shown in Fig 1 and the spillway works redesigned. The rock
at the new location was explored with a limited number of borings; no substantial differences between this and the former
exploration was detected.

As the excavation of the spillway channels progressed, the output of useful rock was low and its improvement
with depth smaller than predicted. Both the useful and excavated volumes of rockfill as a function of time are plotted in
Fig 20; by April 1972, the ratio of the above volumes was 0.41.
In addition, the limestone was rather soft (q_u < 500 kg/cm^2)
and it produced large quantities of fine particles. The breakage of rock fragments caused by four passes of a 13 ton vibratory roller was large, resulting in a mass of gravelly material with a high content of fines. At this stage of the construction it was decided to change the design of the dam by
substituting an important amount of the compacted rockfill
zone for sand and gravel, borrowed from alluvial deposits located from 4 to 7 km downstream of the site. Fig 21 presents
the maximum cross-section of the dam as finally built; the to-

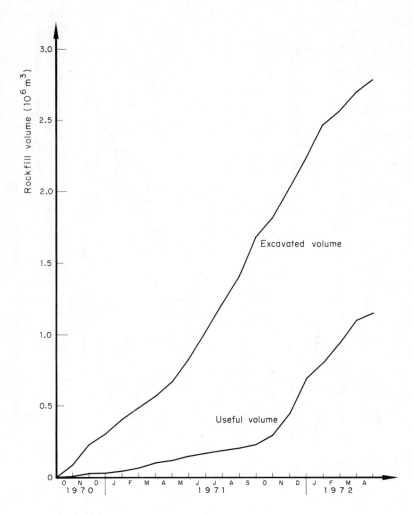

Fig 2O. Useful and excavated volumes of rockfill obtained
from the spillway

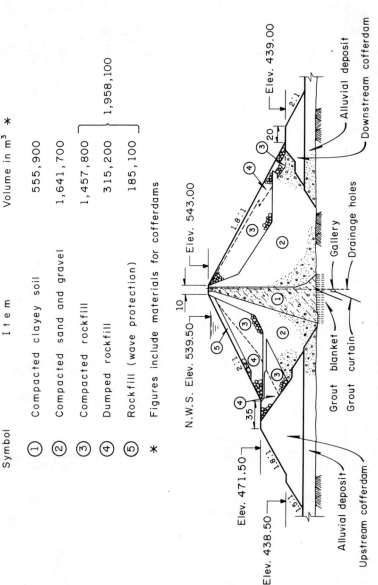

Symbol	Item	Volume in m³ *
①	Compacted clayey soil	555,900
②	Compacted sand and gravel	1,641,700
③	Compacted rockfill	1,457,800
④	Dumped rockfill	315,200 ⎫ 1,958,100
⑤	Rockfill (wave protection)	185,100 ⎭

* Figures include materials for cofferdams

Fig 21. Maximum cross-section of the dam as built

tal volume of the rockfill placed was 1.96 x 10^6 m^3, 54 per cent of the material excavated at the spillway area. As a consequence, the actual cost of the dam compared with its bidding estimate increased 8 per cent.

The use of the gravelly materials in lieu of the rockfill was beneficial. Due to the poor quality of the rock obtained from the reefy limestone, the deformation of the pervious mass induced by grain breakage would have been large, particularly during the first filling of the reservoir, and their effect upon the core almost unpredictable.

COMMENTS

a) Two problems in La Angostura Project remain uncertain, i.e., the permeability of the reservoir and the rate of seepage losses through the abutments. The first filling of the reservoir by the end of 1974 will allow us to confirm or rectify 1) the conclusion about the permeability based on the piezometric observations in the reservoir, and 2) the prediction about seepage losses through the fractures in both banks.

b) The reefy limestone in the spillway area, not detected until the excavation for the discharge channels was well advanced, brought substantial changes in the construction of the dam. Its original design contemplated the use of a large quantity of rock from the spillway. Due to the poor quality of the excavation product, the pervious zones of the dam were built with sand and gravel protected with the best rockfill

obtained from the abovementioned source. One has to admit
that exploration was limited and that our experience of this
type of sedimentary formation was scanty.

   c) The geotechnical studies at the site not only contributed to the design of the underground powerhouse and other
appurtenant structures, but also provided a better understanding of the tectonics and the behavior of the rock mass upon
several types of excavation. Furthermore, the measurements in
the spillway and Diversion Tunnel 1 suggested the possible
causes for the differences between observed and computed displacements in the rock around the cavern for the powerhouse.

REFERENCES

1. Cruickshank C., 1970, "Flujo en un sistema de fracturas
   verticales", Publ. of the Instituto de Ingeniería, UNAM,
   N° 263

2. Covarrubias S., 1970, "Análisis de esfuerzos en la presa
   La Angostura", Report of the Instituto de Ingeniería, UNAM
   to CFE

3. Ramírez de Arellano L. and Moreno E., 1972, "Field Measurements at La Angostura Cofferdams", ASCE Specialty Conference, Purdue University, Vol. I, Part 1

4. Alberro J., 1973, "La Angostura Dam Underground Powerhouse:
   Prediction and Measurement of Displacements during Excava-
   tion", Third International Conference on Rock Mechanics,
   Denver, Col.

5. Figueroa J., 1973, "Sismicidad en Chiapas", Publ. of the
   Instituto de Ingeniería, UNAM, N° 316

SUMMARY

by Edward D. Graf*

From the many cases of pressure grouting solutions to dam foundation problems presented at this conference, it stands out that European practice is far advanced over American practice, both in materials and techniques. The reason for this is that major dam foundations in the United States are grouted in accordance with the "cook book" specifications of either the Corps of Engineers or the Bureau of Reclamation which are, essentially, the same specifications which were in effect 40 years ago.

American practice is to have the grouting accomplished by the low bidder (labor and equipment broker) which, for a number of reasons, leaves little alternative but continuation of the portland cement and water practices developed 50 years ago. American grouting engineers experienced in this system are extremely knowledgeable in the techniques of using cement and water. A European grouting engineer has a thorough knowledge of the wide range of materials available and the various techniques to success- fully inject these materials.

In the United States there are a handful of grouting companies that use the wide range of materials and techniques available. These companies operate as engineering services in the European style of including both design and construction. However, these companies do not grout major dams in this country because their engineering service approach does not fit the standard specification system. Occasionally, small dams in the United States are grouted using the available modern technology.
_____
*President, Pressure Grout Co., Daly City, Calif.

465

Modern grouting, like site exploration, requires intelligent and experienced field personnel with the knowledge and ability to make immediate changes in materials and techniques as the work progresses. It does not lend itself to "cook book" operations.

The technology and experience for effective and economic grouting is available in the United States. It will not be available for heavy dam construction until grouting is performed as an engineering service.

Subject Index

467

Author Index